Wissenschaftsethik und Technikfolgenbeurteilung
Band 6

Schriftenreihe der Europäischen Akademie zur Erforschung
von Folgen wissenschaftlich-technischer Entwicklungen
Bad Neuenahr-Ahrweiler GmbH
herausgegeben von Carl Friedrich Gethmann

K.-M. Nigge

Life Cycle Assessment of Natural Gas Vehicles

Development and Application of Site-Dependent
Impact Indicators

With 30 Figures and 44 Tables

 Springer

Reihenherausgeber
Professor Dr. Carl Friedrich Gethmann
Europäische Akademie zur Erforschung von
Folgen wissenschaftlich-technischer Entwicklungen
Bad Neuenahr-Ahrweiler GmbH
Wilhelmstraße 56, D-53474 Bad Neuenahr-Ahrweiler

Autor
Dipl.-Phys. Karl-Michael Nigge M.E.S.
Rheinweg 27
D-53113 Bonn

Redaktion
Dagmar Uhl. M.A.
Europäische Akademie GmbH
Wilhelmstraße 56, D-53474 Bad Neuenahr-Ahrweiler

Zugl.: Clausthal, Techn. Univ., Diss., 2000

D 104

ISBN-13: 978-3-642-64121-3   e-ISBN-13: 978-3-642-59775-6
DOI: 10.1007/978-3-642-59775-6

CIP data applied for

Springer-Verlag is a company in the BertelsmannSpringer publishing group

© Springer-Verlag Berlin Heidelberg 2000
Softcover reprint of the hardcover 1st edition 2000

Typesetting: Camera-ready by author
Cover layout: de'blik, Berlin
SPIN: 10764216   Printed on acid-free paper   62 / 3020 hu - 5 4 3 2 1 0

European Academy

for the Study of Consequences
of Scientific and Technological Advance

Bad Neuenahr-Ahrweiler GmbH

The European Academy

The *European Academy* is concerned with the scientific study of consequences of scientific and technological advance for the individual and social life and for the natural environment. The European Academy intends to contribute to a rational way of society of dealing with the consequences of scientific and technological developments. This aim is mainly realised in the development of recommendations for options to act, from the point of view of long-term societal acceptance. The work of the European Academy mostly takes place in temporary interdisciplinary project groups, whose members are recognized scientists from European universities. Overarching issues, e.g. from the fields of Technology Assessment or Ethics of Science, are dealt with by staff of the European Academy.

The Series

The Series „Ethics of Science and Technology Assessment" serves to publish the results of the European Academy's work. It is published by the Academy's director. Besides the final results of the project groups the series includes volumes on general questions of ethics of science and technology assessment as well as other monographic studies.

Foreword

In the context of conducting research on the consequences of scientific and technological advance, the Europäische Akademie is also concerned with the support of scientists in the doctoral or post-doctoral phase who are working on topics or methods within its research spectrum. The first dissertation supported by the Europäische Akademie is published in this volume of the book series „Wissenschaftsethik und Technikfolgenbeurteilung".

One of the research areas of the Europäische Akademie is the scientific investigation of environmental consequences of new technologies. Energy conversion and transportation are thereby considered as important areas of technological advance. The dissertation follows this thread by comparing the impacts of natural gas vehicles on human health and the environment with those of reference vehicles fueled by petrol and Diesel.

This question is addressed within the framework of Life Cycle Assessment, which is one important instrument of environmental Technology Assessment. Within this framework, a new method for the assessment of impacts on human health is developed and applied. In this way, the dissertation contributes to the methodological research of the Europäische Akademie in the field of Technology Assessment.

The book is addressed to researchers in the fields of alternative fuels, Technology Assessment, and Life Cycle Assessment in particular. It may also be of interest to decisionmakers and the wider public concerned with environmental impacts of energy conversion and transportation. It was written in English in order to be accessible to an international audience.

Bad Neuenahr-Ahrweiler, November 1999   Carl Friedrich Gethmann

Preface

The young discipline of Technology Assessment (TA) provides the basis for an operationalization of the diffuse vision of sustainable development. This assertion implies a need for research in various fields. From the perspective of the natural and the engineering sciences, the following areas of research are concerned:

- description of the present state by means of sustainable development indicators
- ways to deal with imprecise, uncertain and lacking knowledge
- further development of instruments and methods
- normative orientations and ways to deal with normative conflicts
- simulation of dynamic systems.

The dissertation of Karl-Michael Nigge, which was presented to the Faculty of Mining, Metallurgy and Mechanical Engineering at the Technical University of Clausthal, essentially relates to the aspect of the further development of instruments and methods. Within the TA concept, a clear distinction should be made between instruments and methods. Instruments, such as Life Cycle Assessment, are related to a well-defined context of application. Methods, on the other hand, are procedures that can be applied in various contexts.

Methods can be classified in a variety of ways. In terms of their origin, most of the methods come from economics, some of them from the technical sciences and others from military science. With regards to their purpose, analysis, prognosis, evaluation and decision can be distinguished. A third distinction can be made between qualitative and quantitative methods. Utility analysis may be used an illustrative example. It originated in economics and the technical sciences, it can be used for prognosis, evaluation and decision, and it is both qualitative and quantitative in character. Methods can also be characterized as either intuitive-heuristic, statistical, as being based on tree structures or matrices or as model simulations. The latter represent a powerful tool which is often used in connection with scenarios.

Only the practical experience of conducting TA studies can provide a clear view of what is methodologically reasonable. However, Ropohl (1997) identifies a lack of methodological awareness among TA practitioners, especially with regards to the further development of methods. As the main underlying reason, he identifies the circumstance that TA studies are mostly conducted under high time-pressure in the form of projects outside of universities. This deficit represents an opportunity for university based research, which we are pursuing at the Institute of Applied Mechanics at the Technical University of Clausthal.

The dissertation of Karl-Michael Nigge represents a very successful contribution to the development of methods for Life Cycle Assessment. This is achieved in the context of a topic of highest practical relevance, namely the comparison of natural gas, petrol and Diesel as fuels for road vehicles.

I thank my colleague, Professor Dr.-Ing. Hans-Peter Beck, Head of the Institute of Electrical Power Engineering at the Technical University of Clausthal, for stimulating discussions in the context of this dissertation and for kindly accepting to be the second referee.

Clausthal-Zellerfeld, January 2000   Michael F. Jischa

Contents

Abbreviations and Symbols[1]

φ	angular variable in cylindrical coordinates
ρ	population density
ρ_{eff}	effective population density
σ	dispersion parameter in Gaussian dispersion model
σ_g	geometric standard deviation of a lognormal distribution
Γ	Gamma-function
τ_a	atmospheric residence time
a (variable)	scale-parameter of the Weibull distribution
a (unit)	year
b	background pollutant concentration
B[a]P	benzo[a]pyrene
c	pollutant concentration
C	settlement structure class
carc.	carcinogenic
CED	Cumulative Energy Demand
CNG	Compressed Natural Gas
D (variable)	damage
D (index)	Germany
DALY	Disability Adjusted Life Years
DIS	Draft International Standard
E	effect factor
EE	end energy
EIA	Environmental Impact Assessment
EMAS	Environmental Management and Auditing Scheme
EMEP	European Monitoring and Evaluation Program
ERF	exposure response function
EU	European Union
EURO	European emission control standard
F	fate and exposure factor
GEMIS	Gesamt-Emissionsmodell Integrierter Systeme
GJ	GigaJoule
GWP	Global Warming Potential
h	effective emission height
H	height of the atmospheric mixing layer
H_u	lower calorific value
i	index for emission sites

---

[1] Organizations featuring as authors of quoted literature are referred to by their acronyms in the text. Their full names can be found in the list of references.

I	population exposure per mass of emitted pollutant (PE/M)
IARC	International Agency for Research on Cancer
IPCC	Intergovernmental Panel on Climate Change
ISO	International Standards Organization
k	shape-parameter of the Weibull distribution
L	driving distance
LCA	Life Cycle Assessment
LCI	Life Cycle Inventory
LCIA	Life Cycle Impact Assessment
M	mass of emitted pollutant
MIR	Maximum Incremental Reactivity
MJ	MegaJoule
NGV	Natural Gas Vehicle
NL	The Netherlands
NMVOC	non-methane volatile organic compound(s)
NOAEL	No Observed Adverse Effect Level
NOR	Norway
NUTS	Nomenclature des Unités Territoriales Statistiques
OPEC	Organization of the Petroleum Exporting Countries
p, pers.	persons
PAH	polycyclic aromatic hydrocarbons
PE	population exposure
PM	particulate matter
PM 2,5	particulate matter with diameter $< 2,5\ \mu m$
PM 10	particulate matter with diameter $< 10\ \mu m$
POCP	Photochemical Ozone Creation Potential
Q	emission rate
r, R	radius
RA	Risk Assessment
resp.	respiratory
RUS	Russia
SETAC	Society of Environmental Toxicology and Chemistry
t	time
T	duration of an emission
TA	Technology Assessment
TJ	TeraJoule
u	wind speed
v_d	dry deposition velocity
VOC	volatile organic compound(s)
v_w	wet deposition velocity
w	relative frequency of wind directions
WTM	Windrose Trajectory Model
$\underline{x}$	2-dimensional position vector for location on the surface of the Earth
YLD	Years Lived Disabled
YLL	Years of Life Lost
z	height above ground

1 Introduction

Since the late 1960s, environmental problems such as the depletion of finite resources, the pollution of air, water and soil, global climate change and stratospheric ozone depletion caused by human activities like energy conversion, transport, agricultural and industrial production, forestry, tourism and private consumption have become a major concern worldwide. Attempts to solve them can pursue three main strategies:

- limiting population growth
- reducing the per capita consumption of products and services (sufficiency strategy)
- providing the same amount of products or services with lower environmental burdens (efficiency strategy).

In most industrialized countries, population growth is not an issue of concern. This leaves a combination of sufficiency and efficiency strategy to address their environmental problems. The present work stands in the context of the strategy of increasing the environmental efficiency in the area of transportation.

In pursuing this strategy, trade-offs frequently occur between the different stages of the life cycle of products or services. The use of automobile parts made from aluminum or magnesium instead of steel, for example, reduces the vehicle weight and therefore the fuel consumption, but their production requires a higher energy input than in the case of steel. Trade-offs may also occur between different types of environmental impacts. Natural gas vehicles, for example, cause lower impacts on human health than vehicles fueled by petrol or Diesel, but due to their higher fuel consumption, they require more fossil resources to be extracted and may also have disadvantages in terms of climate change.

Life Cycle Assessment is a tool to analyze and assess these trade-offs by taking into account all relevant stages of the life cycle of a product or service 'from cradle to grave' and by considering a broad spectrum of environmental impacts. In a Life Cycle Assessment, an inventory of relevant inputs and outputs (e.g. resources and emissions) of the numerous processes within the life cycle of a product or service is compiled. They are assessed with regards to environmental impacts such as resource depletion and damage to human health or ecosystems.

A deficit of the standard methods used for the assessment of environmental impacts within Life Cycle Assessment is that they are frequently based on unrealistic simplifications. For example, the impact of an emission of a toxic airborne pollutant on human health is considered to be the same no matter whether it occurs in a large city or in a sparsely populated area. For pollutants with atmospheric residence times in the order of months or years, which disperse globally, this assumption is indeed reasonable. In the case of pollutants with shorter atmospheric residence times in the order of hours or days, however, it disregards the fact that a much higher number of people is exposed to the pollutant when it is emitted in a city.

In order to quantify this dependence of the exposure of the population on the site of the emission, numerous dispersion models for air pollutants are available in principle. They are routinely applied to consider airborne pollutant emissions from individual industrial facilities or processes. Within Life Cycle Assessment, however, a large number of emission processes needs to be considered. The collection of the input data required by the available pollutant dispersion models for this large number of processes would require too much time and effort.

This raises the question whether existing pollutant dispersion models can be simplified in such a way as to be applicable with reasonable effort within Life Cycle Assessments. A new method is presented here (chapter 3) which addresses this problem for the case of health effects of airborne pollutants from transportation and energy generation. The method provides site-dependent indicators which allow to differentiate the health impacts of these airborne pollutants between emissions in cities of various sizes or in rural areas.

These indicators are applied within a Life Cycle Assessment of natural gas vehicles compared to vehicles fueled by Diesel and petrol (chapters 4 and 5). The purpose of the use of natural gas as an alternative fuel is the reduction of emissions of airborne pollutants which affect human health. Natural gas vehicles are potentially becoming more widespread in many countries. However, a sufficiently large network of fuel stations providing compressed natural gas is still lacking. Therefore, the first step in the market introduction of natural gas vehicles is their use in vehicle fleets such as city buses, garbage collection trucks, delivery vehicles or taxis. Due to their defined location of use, they can obtain the gas from a few central fuel stations.

The location of the use of these vehicles is also of interest with regards to the health benefits that can be achieved by using natural gas instead of Diesel or petrol. It is frequently supposed that the effort associated with the switch to natural gas is only worthwhile for vehicles driving in large cities. However, due to the dispersion of the air pollutants emitted from the vehicles, significant contributions to the overall health benefits may also come from areas up to several hundred kilometers away. The extent to which the health benefits of using natural gas vehicles instead of Diesel or petrol vehicles differ between cities of various sizes therefore needs to be quantified.

This is achieved by using site-dependent impact indicators calculated according to the method presented here. The method thereby allows to treat the vehicle emissions and the numerous 'upstream' emissions during the extraction, processing and transport of the fuels in a consistent way, i.e. with the same level of detail. This is particularly important for the natural gas vehicles since reductions in the vehicle emissions increase the relative importance of the upstream emissions.

2 Life Cycle Assessment

In this chapter, Life Cycle Assessment is located within the wider context of Technology Assessment. A definition of Life Cycle Assessment and its phases is provided. Methods used in each of the phases are briefly described. Methods for the assessment of human health effects of air pollutants are discussed in detail in order to provide a background for the new method presented in chapter 3.

2.1
Life Cycle Assessment as an Instrument of Technology Assessment

The concept of Technology Assessment (TA) emerged in the late 1960s in the United States. Its central idea is to evaluate existing and, in particular, new technologies not only according to their technical functionality and micro-economic feasibility, but also to take into account possible macro-economic, social or environmental effects. The Guideline 3780 „Technology Assessment - concepts and foundations" of the German engineering society Verein Deutscher Ingenieure (VDI) defines Technology Assessment (TA) as

> the planned, systematic, organised procedure that
> - analyzes the state of a technology and its possible developments
> - estimates direct and indirect technical, economic, health-related, ecological, human, social and other consequences of this technology and possible alternatives
> - assesses these consequences based on defined goals and values or also demands further desirable developments
> - develops and elaborates upon possible courses of action on this basis
> (VDI 1991:2, translation by the author).

While these cognitive goals of TA are widely accepted, a variety of different approaches have been developed with regards to their realization. A recent review is provided by Grunwald (1999). A particular point of dissent concerns the treatment of normative choices required for the assessment of the consequences of new technologies. Options in this regard are the provision of a set of scenarios characterized by different normative presuppositions, the use of participatory methods to ensure the acceptance of the presuppositions or the investigation of their acceptability by means of philosophical ethics (Gethmann 1999).

It is clear that the ambitious cognitive goals of TA can hardly be simultaneously and completely fulfilled in practice[1]. In particular, it is rarely the case that the en-

---

[1] Overviews of methods of TA provided by various authors (Huisinga 1985; VDI 1991: 16-19; Ludwig 1995: 56-66; Ropohl 1997; Steinmüller 1999), in addition to containing rather general

tire spectrum of possible consequences of an emerging technology is covered in one study. Instead, the actual practice of TA has to a large extent focused on environmental problems (Kreibich 1999: 813-814).

Technology Assessment has been institutionalized in various industrialized countries in order to provide advice on new technologies to political decisionmakers in the legislature or the executive at the regional, national or international level. This development was initiated in 1972 by the establishment of the Office of Technology Assessment of the Congress of the United States of America, which was operating until 1995. Subsequent institutionalizations of TA as policy analysis mainly occurred in Europe (see Bröchler et al. 1999:vol. 2 for an overview).

Parallel to and for the most part institutionally independent from the development and use of TA for the purpose of policy analysis, a number of instruments have been developed and are routinely used to address more standardized types of *environmental* issues within regulatory and economic contexts of decisionmaking. Among these instruments are Risk Assessment, Environmental Impact Assessment, Eco-Audit and Life Cycle Assessment.

Risk Assessments (RA) address the risks to human health and the environment posed by industrial activities either during normal operation or as a result of accidents. They are typically used in the regulation of new and existing chemicals (e.g. agrochemicals, food additives) or of technical facilities (e.g. nuclear power plants). Environmental Impact Assessments (EIA) are often required for the approval of new infrastructure projects (e.g. streets or waste disposal sites) and typically focus on the effects of such installations on natural ecosystems. Eco-Audits may be voluntarily used by businesses and involve an inventory of the energy and material inputs and outputs of the individual production sites. Life Cycle Assessments (LCA) of products or services are also used on a voluntary basis mostly by enterprises. A more detailed overview of these instruments is provided by Ludwig (1995:35-44, 49-51).

Despite their institutionally independent development, they can logically be seen as operationalizations of specific aspects of the overarching concept of TA. They represent partial TA studies in the sense that they may be limited to particular types of technical systems and impacts (mostly environmental) and may only address the status quo as opposed to also addressing possible future developments and courses of action (Ropohl 1999:35). Therefore, they shall be denoted as *instruments of Technology Assessment* here (Ludwig 1995; Jischa 1997, 1999a, b)[2].

Due to the more standardized problems they address, their procedures are also more standardized than it is possible in the case of TA as policy analysis. The methodological framework of Life Cycle Assessment, for example, is currently being standardized within the ISO 14040-14043 norms of the International Stan-

---

procedural schemes of problem solving, therefore inevitably have the character of somewhat coincidental lists of possible methods drawn from a wide variety of scientific disciplines rather than that of an overall coherent methodology of TA.

[2] The denotation of Life Cycle Assessments etc. as instruments refers to their *results*, which can be used within TA or for the purpose of decisionmaking. At the same time, the *procedure* of arriving at these results is a systematic combination of various methods and can therefore be denoted as a method itself. Method is thereby understood as a procedure which is planned with regards to purpose and means (Mittelstraß 1995:876).

dards Organization (ISO 1997, 1998, 1999a, b)[3]. In a similar way, the principles of Risk Assessments of existing and new chemicals are described in Directives and technical guidance documents of the European Commission (van der Wielen 1996:3). Despite the standardization of certain elements of these instruments, further methodological improvements are a topic of intensive current research. This provides a counterexample to the lack of the use and improvement of methods of Technology Assessment that is sometimes deplored (Zimmermann 1993; Ropohl 1997:196-197, 1999:34-35).

2.2
Definition and Phases

Since 1994, Life Cycle Assessment has been the topic of ongoing standardization efforts within the International Standards Organization (ISO). Several norms concern the definition of LCA as well as the individual phases within an LCA. The standards ISO 14040 on the principles and framework of LCA and ISO 14041 on goal and scope definition and life cycle inventory analysis have been adopted in June 1997 and September 1998, respectively. ISO/DIS 14042 on Life Cycle Impact Assessment and ISO/DIS 14043 on Life Cycle Interpretation currently hold the status of Draft International Standards (DIS) (Curran 1999:123). Accordingly, Life Cycle Assessment is defined as a

> compilation and evaluation of the inputs, outputs and the potential environmental impacts of a product system throughout its life cycle (ISO 1997:5).

This definition is expanded upon in the following way:

> LCA is a technique for assessing the environmental aspects and potential impacts associated with a product, by:
> - compiling an inventory of relevant inputs and outputs of a product system;
> - evaluating the potential environmental impacts associated with those inputs and outputs;
> - interpreting the results of the inventory analysis and impact assessment phases in relation to the objectives of the study.
> LCA studies the environmental aspects and potential impacts throughout a product's life (i.e. cradle-to-grave) from raw material acquisition through production, use and disposal. The general categories of environmental impacts needing consideration include resource use, human health, and ecological consequences (ISO 1997:3).

Accordingly, what distinguishes LCA from other instruments of environmental analysis and management is the striving for comprehensiveness in two respects:

1. LCA is not limited to individual processes, industrial installations or projects, like Risk Assessment and Environmental Impact Assessment are. Instead, all processes connected with a product or service ‚from cradle to grave', i.e. from the extraction of raw materials to the disposal or recycling of the products, are to be considered. This ensemble of processes is denoted as product system. The

---

[3] ISO 14040 and 14041 have been put in place, while the remaining norms currently hold the status of Draft International Standards.

smallest parts of the product systems for which input and output data are collected are called unit processes.

2. LCA is not limited to particular impacts such as climate change or carcinogenicity, but is to cover an encompassing spectrum of impact categories which operationalize the general areas of resource use, human health and ecological consequences.

The motivation underlying this comprehensive approach is to avoid that trade-offs between either the different life cycle phases (1.) or between the different impact categories (2.) are overlooked, i.e. to avoid that impacts are merely shifted from one phase or category to another and are eventually even increased thereby. A trade-off of the first type occurs, e.g., between the reduction of the fuel consumption of vehicles through the use of aluminum as a light-weight material instead of steel and the increased energy required for its production (see, e.g., Saur et al. 1996). A trade-off of the second type occurs, for example, in the case of replacing Diesel vehicles by natural gas vehicles. While natural gas vehicles perform better in terms of human health effects, they have disadvantages in terms of the extraction of fossil resources and, depending on their fuel consumption, possibly also in terms of climate change (see case study in chapters 4 and 5). In the case of trade-offs of the second type, the overall preference for one of two product alternatives depends on the relative importance that the decisionmakers assign to the different impact categories (see section 2.5.1).

One important notion in the context of LCA is the term ‚functional unit', which is defined as a „quantified performance of a product system for use as reference unit in a life cycle assessment study" (ISO 1997:3). All environmental inputs and outputs of the considered product systems are normalized to the functional unit. In the case of a comparison of two or more product systems or services, this ensures that the same utility is considered in each case. In the case of packaging materials, for example, it does not necessarily make sense to compare 1 kg of material A and 1 kg of material B. Rather, the masses m_A and m_B of the materials required for the packaging of an equal amount of a specified product over the same time and keeping the product in the same quality etc. are to be compared. Likewise, in the case of comparing means of transportation of persons, the underlying service of interest is not the movement of the vehicles over a specified distance, but the transportation of a specified number of persons. Therefore, the functional unit has to be expressed in person-kilometers, which can only be reduced to vehicle-kilometers if the vehicles offer the same space and can be expected to operate with the same occupancy rate[4].

According to ISO 14040, an LCA study shall include the following phases:

1. Goal and Scope Definition
2. Inventory Analysis

---

[4] Of course, it is unlikely that two products or services are identical in all of their functional parameters. In the case of transportation means, for example, other parameters besides the person-kilometers, which may be of interest for the definition of the functional unit, could be range, safety, peak velocity, acceleration and maintenance intervals, just to name a few. The selection of certain functional parameters as being part of the functional unit depends on the goals of the LCA study.

3. Impact Assessment
4. Interpretation

Their definitions and related terminologies and methods will be briefly discussed in the following. Only the Life Cycle Impact Assessment of human health effects of airborne pollutants will be considered in more detail to provide the background for the method presented in chapter 3. Furthermore, characterization factors[5] for the assessment of impacts in the categories of relevance to the case study (chapters 4 and 5) will be provided.

2.3
Goal and Scope Definition

The standards ISO 14040 and 14041 require the goal and the scope of an LCA study to be made explicit. The goal comprises the intended application, the reasons why the study was conducted and the intended audience. In the definition of the scope, the following points shall be addressed:

1. the function of the product systems
2. the functional unit
3. the product system to be studied
4. the product system boundaries (i.e. the delimitation of the processes that are included in the study)
5. procedures for the allocation of environmental burdens in the case of multi-output processes
6. types of impacts and methodologies of impact assessment and subsequent interpretation to be used
7. data requirements
8. assumptions
9. limitations
10. initial data quality requirements, if any
11. type and format of the report required for the study.

2.4
Inventory Analysis

In the inventory analysis phase, data are collected and calculations are carried out in order to quantify the relevant inputs and outputs of the product systems under consideration (ISO 1997:9). The encompassing approach of LCA generally puts high demands on the data collection. Databases for certain sets of standard processes (e.g. energy systems, production of frequently used materials, transport systems) are continuously being compiled and typically included in various commercial LCA software packages that facilitate the tasks of data collection, data administration and inventory calculation. A few databases are publicly accessible, among

---

[5] see definition in section 2.5.1.

them the database GEMIS (Gesamt-Emissionsmodell Integrierter Systeme) for energy systems (Rausch et al. 1998) used for the case study in chapters 4 and 5. Nevertheless, the ideal that all inputs and outputs at the system boundaries are elementary flows (i.e. flows with no prior or subsequent human transformations; ISO 1997:4; ISO 1998:10) can hardly be achieved in the definition of the system boundaries. Instead, processes that are estimated to be insignificant for the overall results, e.g. on the basis of proxy indicators such as energy or mass or previous experience with similar product systems, are usually cut off in practice. Iterations involving modifications of the initially defined system boundaries may hence be required.

With regards to data collection, two approaches can be distinguished. The approach of process chain analysis rests on the collection of detailed input and output data for individual processes. In contrast to this micro-level analysis, economic input-output analyses provide data at a macro-level which are aggregated over groups of processes or industry sectors (Hendrickson et al. 1998). While such input-output tables traditionally only contain economic data, several national statistics are now supplemented by environmentally relevant data such as energy consumptions or emissions. The two approaches can be seen as complementary, with input-output analysis supplementing data for those processes for which a micro-level analysis requires too many resources in relation to their expected significance (VDI 1997:13; Marheineke 1998).

Apart from the practical problems of data collection under the conditions of limited resources, a more fundamental methodological problem arises with regards to the allocation of environmental burdens of processes with multiple products, which are not all related to the functional unit under investigation. A typical example is the refining of petroleum, which simultaneously yields a variety of products such as petrol and Diesel fuel, while only one of the products is of related to the functional unit, e.g. Diesel fuel to 1 kilometer of driving with a Diesel bus (see section 4.2.2.1).

ISO 14041 establishes a stepwise procedure to deal with this problem. In the first step, the possibility of avoiding an allocation should be examined, either by dividing the considered unit processes into subprocesses with separate outputs, in which case the co-production was only an artifact of insufficiently detailed data collection, or by redefining the functional unit so as to include the additional functions of the co-products, which may lead to a significant increase of the amount of data to be collected. If allocation cannot be avoided, it should preferably reflect the actual physical way in which inputs and outputs are changed by changes in the products or functions of the process. Only if such physical relationships cannot be established to a sufficient extent should allocation be based on other measures such as the economic value of the co-products (ISO 1998:17-18). Similar allocation procedures are also required to deal with recycling processes, which shall not be discussed here in detail, since it is not relevant to the case study (see ISO 1998:18-20).

Once the data collection for the inputs and outputs of the unit processes has been completed, the inputs and outputs of the entire product system, normalized to the functional unit, need to be calculated. If the product system consists of n+1 unit processes (one of them being the functional unit) the task is to determine a scaling factor for each of n unit processes that corresponds to the demand of (usually) one particular input or output imposed on the process directly or indirectly by the functional unit.

In a representation of the product system by a flow diagram, the n+1 unit processes are connected by n lines representing flows from a particular output of one unit process to a particular input of another unit process. In the easiest case, the flow diagram describing the product system has a tree structure, i.e. there are no junctions between the flows and no loops. In that case, the scaling factors can be determined by the propagation method, i.e. starting from one (suitably chosen) unit process, they can subsequently be determined from equations with one unknown. If the flow diagram contains junctions or loops, the n unknown scaling factors generally need to be determined by solving a set of n linear equations (Schweimer 1998). This can either be done by matrix operations or by iterative calculations. The GEMIS software (Rausch et al. 1998) used in the case study solves the linear equations approximately by an iterative application of the propagation method until a preselected degree of convergence has been reached (Fritsche et al. 1994:224).

2.5
Impact Assessment

2.5.1
General Framework and Impact Categories

The following terminology and definitions applying to the impact assessment phase of LCA are based on the Draft International Standard ISO/DIS 14042 on Life Cycle Impact Assessment and on the background document for the Second Working Group on Life Cycle Impact Assessment of the Society of Environmental Toxicology and Chemistry (SETAC) Europe (Udo de Haes et al. 1998). In cases where the two sources use different terms, the source is briefly indicated in brackets as either ISO or SETAC.

The purpose of the impact assessment phase is to relate the results of the inventory analysis, i.e. a comprehensive list of energy and other resource demands, emissions to air, water and soil and other *environmental interventions* (SETAC) for each of the considered options for the realization of the functional unit to a number of environmental issues of concern. According to the encompassing approach of Life Cycle Assessment, a broad spectrum of consequences of the environmental interventions contained in the inventory is to be considered, corresponding to a wide range of pollutants and other interventions to be contained in the inventory. The spectrum of possible impacts can be organized at a general level into several *safeguard subjects* (SETAC) or *areas of protection*. While the ISO definition quoted above (section 2.2) lists three areas of protection, namely human health, natural resources and natural environment, it has been suggested to add man made environment as a fourth safeguard subject to consider, for example, the damages to buildings by acidifying pollutants (Udo de Haes et al. 1999).

In order for impact assessment to be operational, the safeguard subjects are differentiated into impact categories. While impact categories are generally more detailed than safeguard subjects, one impact category can nevertheless be related to more than one safeguard subject. Several default lists of impact categories to be considered in Life Cycle Impact Assessment have been suggested in the past (e.g. Heijungs et al. 1992:42; Consoli et al 1993:24; Udo de Haes 1996:19). Despite some differences between the various proposals, there is also a high degree of similarity. The ISO/DIS 14042 standard does not contain a list of impact categories, but only formulates a set of general requirements including a grounding on an international agreement or an approval by a competent international body. With this aim, the Second Working Group on Life Cycle Impact Assessment of SETAC-Europe has suggested the following list of impact categories (Udo de Haes et al. 1999):

1. extraction of abiotic resources
2. extraction of biotic resources
3. land use, with the subcategories
   a. increase of land competition
   b. degradation of life support functions
   c. bio-diversity degradation
4. climate change
5. stratospheric ozone depletion
6. human toxicity
7. eco-toxicity
8. photo-oxidant formation
9. acidification
10. nutrification.

This proposal will be followed here, except that only a selection of the most relevant of the above impact categories will be considered in the case study in chapters 4 and 5. It should be borne in mind that impact assessment methods for some categories can be adopted from other fields of work, e.g. in the case of climate change, whereas many methods have been and are developed specifically for the purpose of Life Cycle Impact Assessment (including the method presented in chapter 3 of this work). Therefore, lists of impact categories will likely be subject to change over time due to the availability of new or improved methods. It has also been suggested to introduce indicators for as yet unknown environmental damages (Hofstetter 1998:118-130).

Each impact category is characterized by one or several *endpoints*, which are attributes or aspects of the safeguard subjects, and one or a number of possible cause-and-effect chains associating certain environmental interventions with effects on the category endpoints. The set of cause-and-effect chains is also called the *environmental mechanism* of the impact category (ISO). Variables within the cause-and-effect chains lying between the interventions and the endpoints are called *category midpoints* (SETAC North America).

The qualitative establishment of the relations between interventions and impact categories is called *characterization*. Different interventions may contribute to the same impact category, but the same intervention may also contribute to various impact categories. For example, airborne emissions of both sulfur dioxide (SO_2) and nitrogen oxides (NO_x) contribute to the acidification of soils and lakes, while NO_x also contributes to photo-oxidant formation, and both NO_x and SO_2 also contribute to human toxicity, mostly through the formation of secondary nitrate and sulfate aerosols.

Furthermore, it is desirable to express the contribution of the environmental interventions to the impact categories in quantitative terms. For this purpose, a quantifiable variable within the environmental mechanism of each impact category is defined as *category indicator*. The environmental interventions within each impact category are aggregated on the basis of the category indicator. The factor by which an environmental intervention (e.g. the mass of an emitted pollutant) is to be multiplied to express it in terms of the category indicator is called *characterization factor* (sometimes also equivalency factor).

According to ISO/DIS 14042, the calculation of the category indicators can be based either on empirical relationships or on theoretical models. Due to difficulties within either approach in dealing with the complexities of the environmental mechanisms, the category indicator is often chosen at a midpoint level or sometimes even at the level of the environmental interventions themselves rather than at the level of the endpoints. The degree to which the category indicator can be linked to the category endpoint, either qualitatively or qualitatively, is denoted as its *environmental relevance* (ISO). Because of the difficulties of quantifying the environmental mechanisms, there is usually a trade-off between the ease with which the category indicator can be calculated on the one hand and its environmental relevance on the other hand.

Life Cycle Assessments that are to be used for public comparative assertions, i.e. for public statements of the type that product A is more environmentally friendly than product B, are subject by ISO/DIS 14042 to the stringent requirement that category indicators shall be scientifically and technically valid. In particular, this means that no value choices are allowed in their derivation. In this regard, value choices, such as the weighting of the relative severity of various effects on human health or the weighting of the relative importance of various impact categories, are distinguished from technical assumptions such as an estimate of the atmospheric lifetime of a pollutant for which no measurements are available.

Even though the term ,technical assumptions' is not explicitly defined in ISO/DIS 14042, the underlying distinction seems to be the same as that between explicitly normative issues and conditionally normative issues that has been proposed in the context of Risk Assessment by Brunk et al. (1991). Accordingly, conditionally normative issues could, in principle, be resolved on the basis of empirically validated knowledge, if only such knowledge was available. The function of technical assumptions is therefore to bridge the gaps of knowledge left by scientific uncertainties. In contrast, inherently normative issues can only be resolved by means of endorsing a normative presupposition[6].

Similar to the distinction between Risk Assessment and Risk Management (see Jasanoff 1987, 1990; Gieryn 1995), the more recent distinction by ISO between value choices and technical assumptions in the context of LCIA has also been the topic of some debate (Hertwich and Pease 1998; reply by Marsmann et al. 1999). While current best practices for most impact categories likely fulfil the stringent requirement of avoiding value choices, this does not hold for the weighting of the

---

[6] The description of explicitly and conditionally normative issues was adapted from Nigge (1995, unpublished).

relative severity of various effects on human health as embodied in the concept of Disability Adjusted Life Years (DALY, see section 2.5.2.5), which was developed under the auspices of the World Health Organization. Therefore, the impact category of human toxicity would need to be split up into a large number of separate subcategories dealing with the various types of health effects. As an alternative, the admission of internationally accepted normative presuppositions, for which those embodied in the DALY concept may be a candidate, into Life Cycle Impact Assessment has been suggested by Udo de Haes and Jolliet (1999)[7].

Subsequent to the calculation of the category indicator values for a product system, the set of indicator values and the impact categories themselves can be subject to three different types of further operations, the application of which is optional according to ISO/DIS 14042, in order to facilitate the interpretation of the results. These operations are the normalization of category indicators and the grouping and weighting of impact categories.

The category indicators can be *normalized* to a reference value, e.g. the total or per capita value of the category indicator for a reference area (country, region). Impact categories can be *grouped* into sets, either on a nominal basis (e.g. according to their spatial or temporal range), or on an ordinal scale indicating their order of importance, in which case value choices are involved. The relative significance of the impact categories can furthermore be expressed quantitatively by means of *weighting*. In this case, the (typically normalized) category indicator results are expressed in a common unit by multiplying them with numerical weighting factors, which also offers the possibility to aggregate the set of category indicators into one single indicator. Examples of such indicators are the Swiss Eco-Points (Ahbe et al. 1999), the Eco-Indicators 95 and 98 (Goedkoop 1995; Goedkoop et al. 1998) and monetary units in terms of which environmental effects are expressed in studies on external costs (e.g. European Commission 1995; Friedrich et al. 1998 for energy and transport systems). The usefulness of such single-value scores is a matter of debate, since they inevitably embody normative presuppositions. For this reason, ISO/DIS 14042 does not permit their use in public comparative assertions. More generally, the question is to what extent the normative presuppositions required to determine the relative importance of different impact categories are either generally accepted or can be shown to be acceptable on ethical grounds.

2.5.2
Methods for Selected Impact Categories

In the following, category indicators and methods for their calculation will be briefly described for the impact categories of relevance to the case study in chapters 4 and 5. Methods for the assessment of human health effects from airborne

---

[7] In the present work, use is made of the DALY concept to aggregate different effects on human health (section 2.5.2.5). The disaggregation of the DALY figures into Years of Life Lost (YLL) and Years Live Disabled (YLD) is either directly indicated in the result tables, where convenient, or can be determined from the provided data. A further disaggregation of the YLL and YLD indicators into the dominant individual respiratory health effects is possible based on the data listed in table 2.7.

pollutants will be discussed in some more detail (section 2.5.2.5) to provide the background for the method presented in chapter 3. Unless stated otherwise, the definitions of impact categories and the selection of category indicators are those proposed by Udo de Haes et al. (1999). A summary of the respective scientific backgrounds for the purpose of LCA is provided by Hauschild and Wenzel (1998).

2.5.2.1
Extraction of Abiotic Resources

This is one of three impact categories concerning inputs into the product systems, as opposed to outputs such as emissions. The two remaining input related categories are extraction of biotic resources and land use, which are not discussed here since they are not considered in the case study in chapters 4 and 5.

According to the renewal rate of the abiotic resources in relation to the speed of extraction, a distinction is made between deposits, funds and flows. The renewal rate of deposits (e.g. fossil fuels and mineral ores) is extremely slow, such that they are practically depleted. Funds, such as groundwater, sand and clay, can be depleted as well as they can recover. Flow resources, such as solar energy, wind energy and surface water, cannot be depleted, but in situations where they are limited relative to the demand, competition can arise about their use (Udo de Haes et al. 1999:168).

With regards to the extraction of abiotic resources, the general scheme that the Life Cycle Inventory contains all environmental interventions implies that the extracted amounts of each deposit, fund or flow are listed. Energy carries should, due to difficulties of defining their ‚energy content', i.e. their extractable energy, in an unambiguous way (see below), be listed separately in terms of their mass (Frischknecht et al. 1998). In the impact assessment phase, the potentially large number of different extracted resources can be characterized in terms of a category indicator.

The choice of a suitable category indicator, including the question of whether one overall indicator can deal consistently with deposits, funds and flows at the same time, is currently still a matter of research (Udo de Haes et al. 1999:169), in part because different views exist on the meaning of the safeguard subject of natural resources. Options in this regard include a possible intrinsic value of the resources, their non-substitutability, the decrease of the quality of the remaining deposits and the preservation of the option of their future use (Frischknecht et al. 1998:270). Besides some measure for the energy, exergy or entropy content of the resources, several of the suggested category indicators also take into account economic aspects such as the existing amount of deposits or funds, the annual use of the resources, the replenishment rate of funds and the degree of competition over funds or flows. An overview of possible category indicators is given by Finnveden et al. (1996:41-45). If no indicator can be found to suit deposits, funds and flows at the same time, three subcategories will need to be established to deal with these types of resources separately (Udo de Haes et al. 1999:168).

Since the case study in chapters 4 and 5 deals with the energy conversion chains of fossil fuels (Diesel, petrol and natural gas) for use in vehicles, the depletion of the respective deposits of energy resources (raw oil and natural gas) is of particular interest. Differences between the two fuel conversion chains in this regard can arise either from different conversion losses (e.g. different fuel consumptions of the vehicles) or from different amounts of required auxiliary energy inputs for the extraction, processing and transport of the fuels. To account for possible differences in these regards, the use of the (upper or lower) calorific value of the extracted resources as a category indicator appears to be a suitable choice[8]. The Guideline 4600 of the German engineering society VDI (Verein Deutscher Ingenieure) presents a methodological framework for aggregating the extraction of energetic resources over the whole life cycle of a product on the basis of this indicator. The aggregated result is denoted as the Cumulative Energy Demand (CED) of the product (VDI 1997).

In the VDI Guideline 4600, reference to the extraction of energy carriers is made by using the notion of *primary energy* (or crude energy) as describing the energy content of resources that have not yet been subject to any conversion. As a default, the primary energy is measured in terms of the net calorific value, with the option of using the gross calorific value instead. While the determination of the calorific value of the primary energy is straightforward in the case of energy provided from fossil fuels, the determination of analogous definitions for nuclear energy and renewable energies is more difficult (VDI 1997:7-8; Volkmar 1999). However, since most of the energy inputs in the case study come from fossil sources, these methodological problems shall not be discussed here. For the same reason, the inventory data for the case study provided in chapter 4 will, for the sake of simplicity, not contain the masses of the various primary energy carries (mainly raw oil and natural gas) separately. Only their aggregation in terms of the Cumulative Energy Demand will be listed, even though, strictly speaking, it is part of the impact assessment phase.

2.5.2.2
Climate Change

This category comprises all impacts related to climate change caused by changes in radiative forcing. Besides the natural environment (e.g. displacement of vegetation zones), these impacts may also affect human health (e.g. through shifts in the ranges of diseases) and the man-made environment (e.g. through increased damages due to storms). Modeling of such cause and effect chains up to the endpoint is not yet very far developed, however. The relative degree of radiative forcing is therefore used as a category indicator at a midpoint level and expressed in terms of the Global Warming Potentials (GWPs) provided by the Intergovernmental Panel on Climate Change (IPCC) (Udo de Haes et al. 1999:170).

The GWP of a substance is a measure for its cumulative contribution to radiative forcing over a specified period of time T, typically either 20 years, 100 years

---

[8] It should be noted that the choice of a measure representing the extractable energy provided by a resource (such as the lower or upper heating value) is dependent on the available technologies. For example, in the 1960s and 70s, the condensation enthalpy of the water vapor was not used in fossil fuel power plants due to corrosion problems, and correspondingly the lower (or net) calorific value of the fuels was used as a measure in energy statistics. Today, with this technical possibility being available, the upper calorific value should be used. In the future, use of other reactions (e.g. nuclear reactions) may allow for more energy to be used from the same fuels (Frischknecht et al. 1998).

Table 2.1. Global Warming Potentials (GWP) for methane (CH_4) and nitrous oxide (N_2O) integrated over 20, 100 and 500 years (Houghton et al. 1995:22)

substance	residence time [a]	GWP [kg CO_2 / kg substance] 20 years	100 years	500 years
CO_2	variable	1	1	1
CH_4	12 ± 3	56	21	6,5
N_2O	120	280	310	170

or 500 years. The choice of T is critical, and the suitability of a choice depends on the effects to be assessed and the policy measures to be considered. The GWP is a function of the absorption spectrum of the substance in the infrared, its background concentration (due to saturation effects) and its atmospheric residence time. Besides the direct radiative forcing by the substance itself, indirect effects associated with its influence on the atmospheric chemistry are also considered. In the case of methane, such indirect effects are due to tropospheric ozone production and stratospheric water vapor production.

GWPs are expressed in relative units, with CO_2 serving as a reference substance (GWP = 1). The product of a mass of an emission of a substance and its GWP is denoted as CO_2-equivalent. It indicates the mass of CO_2 that would yield the same cumulative radiative forcing (Hauschild and Wenzel 1998; Houghton et al. 1996:13-22). The GWPs of the substances considered in the case study in chapter 4 and 5 are listed in table 2.1.

2.5.2.3
Acidification

Acidifying substances can cause a large variety of impacts on plants (either directly via the air or indirectly via the soil), animals (fish in freshwater ecosystems) and materials. The ,dead' lakes in Scandinavia are examples of impacts that are in part due to acidification (Stanners and Bourdeau 1995:541). A modeling of all these effects up to their endpoint is beyond the scope of the current state of Life Cycle Assessment. For the time being, the use of the number of released hydrogen cations is therefore proposed as a category indicator (Udo de Haes et al. 1999:172). Furthermore, the focus is typically on the acidification of soils or freshwater ecosystems.

The acidification of soils and freshwater ecosystems can be described as the reduction of the acid neutralizing capacity (buffering capacity) of the system, which eventually leads to a fall of its pH value and a subsequent change in the availability of nutrients or harmful metals. Acidification occurs through the addition or release of hydrogen cations. Since hydrogen (or other) cations are always deposited to a system together with anions, the anions need to be leached out of the system in order for actual acidification to occur. Otherwise, one speaks of potential acidification. In LCA, acidification is usually considered through impact potentials, which represent the sum of actual and potential acidification or, in other words, the theoretical maximum number of hydrogen cations that can be released.

Substances that can contribute to acidification are strong acids such as hydro-chloric acid (HCl) or sulphuric acid (H_2SO_4), acidic anhydrides such as sulfur dioxide (SO_2), sulfur trioxide (SO_3) and nitrogen dioxide (NO_2) as well as ammonia (NH_3), which, despite being a base, releases hydrogen cations subsequent to bacterial mineralization in the soil (Hauschild and Wenzel 1998:160-162). The most important primary pollutants are SO_2, NO_x and NH_3, from which the other substances (except HCl) can be formed as secondary pollutants through chemical reactions in the atmosphere. A common way to express the impact potentials of these substances in the context of Life Cycle Assessment is through the mass of SO_2 that could release the same maximum amount of hydrogen cations. Accordingly, the acidification potential is expressed in SO_2-equivalents in a way similar to the CO_2-equivalents in the case of climate change. The acidification equivalency factor EF_{ac}, which indicates the mass of SO_2-equivalents per mass of a pollutant, is calculated as

$$EF_{ac} = (n/n_{SO_2}) \, (m_{SO_2}/m) = (n/2) \, (64{,}06/m[g/mol]). \tag{2.1}$$

n denotes the maximum number of hydrogen cations that can be released by a molecule of the pollutant and m the molecular weight of the pollutant. The corresponding characterization factors for the acidifying substances considered in the case study are listed in table 2.2. These acidification equivalency factors represent a worst case consideration, since they are related to the maximum theoretically possible reduction of the buffering capacity of ecosystems. The degree to which the maximum possible acidification is actually realized depends on various characteristics of the ecosystems.

Furthermore, only the reduction of the buffering capacity of ecosystems was considered so far, but not the question of whether this actually leads to changes of the pH values and subsequently to damages of the ecosystems once the buffering capacity has been exceeded. This threshold behavior of ecosystems with regards to the amount of deposited acidifying substances is quantified through Critical Loads, which are defined as „quantitative estimate[s] of an exposure to one or more pollutants below which significant harmful effects on specified sensitive elements of the environment do not occur according to present knowledge (Task Force on Mapping 1996:I-1, I-2). Accordingly, the risk for a damage to an ecosystem is assumed to be zero for total deposition onto an ecosystem below its Critical Load, and equal to one above the Critical Load, which does not mean, however, that damages are actually known to occur. The desirable expression of actually occurring damages as a more continuous function of the deposition of acidifying substances is not yet available due to the complexity of ecosystems (Potting et al. 1998:85).

Critical Loads for all forest soils and a number of other terrestrial and aquatic ecosystems across Europe (heathland, grassland, peatland, freshwater) are determined in research programs connected to the Long-Range Transboundary Air Pollution (LRTAP) Convention that was developed under the auspices of the United Nations Economic Commission for Europe (UN ECE). Depositions of sulfur and nitrogen exceed the Critical Loads in many areas of Europe, sometimes by many times the size of the Critical Loads (Stanners and Bourdeau 1995:544). Reviews of the concept of Critical Loads and their determination for European

Table 2.2. Acidification equivalency factors for selected airborne pollutants considered in the case study in chapters 4 and 5 (Hauschild and Wenzel 1998:164)

substance	molecular mass m [g/mol]	maximum number n of released H^+ ions per molecule	equivalency factor EF_{ac} [kg SO_2 / kg substance]
SO_2	64,06	2	1
NO	30,01	1	1,07
NO_2	46,01	1	0,70

ecosystems are provided by Bull (1995), Hettelingh et al. (1995) and Posch et al. (1997).

The geographical variation of the Critical Loads leads to a dependence of the acidifying effects of an emission on its location. Spatially differentiated acidification factors for emissions of SO_2, NO_x and NH_3, taking into account the long-range transport and chemical transformation of the pollutants and their geographical pattern of deposition in relation to the spatial distribution of the Critical Loads of ecosystems, were calculated by Potting et al. (1998) on the basis of the RAINS model of acidification (Alcamo et al. 1991). The underlying atmospheric transport model is a Lagrangian trajectory model which was developed by the European Monitoring and Evaluation Program (EMEP) under the UN ECE convention (EMEP/MSC-W 1998). The acidification factors are expressed as the area of the ecosystems which become unprotected (in the sense of an exceedance of their Critical Load) as a result of the considered marginal emission, e.g. in units of hectare per ton. The category indicator is therefore shifted from the release of hydrogen cations more towards the endpoints, without being identical to the endpoints. Due to the change of the background levels of deposition over time because of emission reductions, two sets of acidification factors are provided for the years 1990 and 2010, respectively. The location of the emission is specified in terms of countries.

The acidification factors calculated by Potting et al. (1998) are generally low for emissions in southern and southeastern Europe because of the insensitivity of the receiving ecosystems in combination with the relatively low background emissions in those areas. High acidification factors as a result of sensitive ecosystems apply to emissions from the Scandinavian and Baltic regions and the European part of the former Soviet Union. Emissions in Western and Central Europe are associated with moderate acidification factors since a large number of ecosystems are already unprotected because of the high emissions and depositions in these areas. To what extent this is an artifact of approximating the damage function by a step-function rather than an indication of an actual saturation of damages cannot be determined, however. Emissions at sea cannot generally be disregarded, depending on the prevailing wind direction and the sensitivity and proximity of land areas in that direction. Acidification factors for the Baltic Sea and the North Sea are comparable to those for many countries, while they are very small for the Atlantic Ocean and the Mediterranean. For the regions or countries relevant to the case study, the acidification factors for SO_2 and NO_x according to Potting et al. (1998) are shown in table 2.3. It should be noted that these acidification factors refer to

Table 2.3. Site-dependent acidification factors (marginal changes of the unprotected area) for emissions of SO_2 and NO_x in the years 1990 and 2010 from the countries or regions relevant to the case study in chapters 4 and 5 (Potting et al. 1998:68-71)

country / region	acidification factor SO_2 [ha/ton] 1990	2010	acidification factor NO_x [ha/ton] 1990	2010
Germany (east)	2,17	2,39	0,90	0,87
Germany (west)	1,94	2,32	1,42	1,03
Netherlands	1,24	1,47	0,97	0,88
Norway	10,90	6,87	2,80	1,34
Russia[a]	5,68	0,22	0,89	0,03
Atlantic Ocean	0,19	0,38	0,14	0,22
Mediterranean	0,00	0,00	0,00	0,00
North Sea	1,58	1,83	0,94	0,88
default value[b]	4,51	1,03	1,07	0,26

[a] with the exception of the subregions of Kalingrad, Kola/Karelia and St. Petersburg, to which different acidification factors apply.
[b] area-weighted average of all regions except subregions in Russia and the seas (n = 37, own calculation).

effects on ecosystems as expressed by their Critical Loads. This means that direct effects on plants as well as effects on materials are not included.

2.5.2.4
Nutrification

The impact category of nutrification or nutrient enrichment includes all impacts due to an increased level of macro-nutrients in both terrestrial and aquatic ecosystems. The types of possible impacts are less varied than those of acidifying substances. They concern the terrestrial and aquatic vegetation by way of increased growth of biomass, which may in turn lead to altered species compositions and to oxygen depletion in lakes and coastal waters (Udo de Haes 1999:173). In practice, only nitrogen (N) or phosphorus (P) are relevant limiting factors for the primary production of biomass, such that their addition to an ecosystem can increase the primary production. Therefore, only substances containing N or P in a biologically available form are classified as potential contributors to nutrification. The most significant sources of nutrient enrichment are fertilizers (containing NO_3 and NH_3) used in agriculture, the contents of N and P in waste water from industry and households and the emission of nitrogen oxides (NO_x) from combustion processes in energy production and transportation (Hauschild and Wenzel 1998:186).

Analogous to the calculation of S-equivalents with regards to acidification, nitrogen-equivalents can be calculated by multiplying the mass of an emitted substance with an equivalence factor EF_N for N-enrichment according to

$$EF_N = (\nu/\nu_N)\,(m_N/m) = \nu\,(14,01/m[g/mol]). \tag{2.2}$$

ν denotes the number of nitrogen atoms that can be released by a molecule of the pollutant and m the molecular weight of the pollutant. A characterization factor for P-enrichment can be introduced in the same way. The corresponding equivalency factors for the two substances of interest in the case study (NO and NO_2) are 0,47 g N/g NO and 0,30 g N/g NO_2. The hypothetical worst-case character of these equivalence factors should be noted. They do not take into account which types of ecosystems are reached by the pollutant (terrestrial or aquatic), whether the pollutant can be chemically transformed into a nutrient in the recipient nor whether the recipient is N- or P-limited (Hauschild and Wenzel 1998:186-188). Such differentiations have, for example, been used in a Finnish case study by Seppälä (1999), but are not available in the form of spatially differentiated characterization factors covering emissions across Europe.

2.5.2.5
Human Toxicity

According to the proposal of the Second Working Group on Life Cycle Impact Assessment of SETAC-Europe, the impact category of human toxicity comprises

> all impacts on human health caused by the direct emission of toxic substances both outdoor and indoor, and impacts caused by fine particles and by radiation (Udo de Haes et al. 1999:171).

By focusing on the direct emission of agents (substances or radiation) affecting human health, this impact category is the one that is most directly related to the safeguard subject of human health. In fact, it is the only one exclusively related to that safeguard subject. Many other impact categories also include effects on human health occurring in a somewhat more indirect way: climate change might, for example, shift the geographical ranges of certain diseases, stratospheric ozone depletion is associated with skin cancer due to increased ultraviolet irradiation, and photo-oxidants are secondary pollutants affecting human health.

The distinction as to whether to include a particular health effect in the category of human toxicity is made on a pragmatic basis, considering in particular that the impacts within one category can be captured with one or only a few types of models. A particular question of delimitation arises with regards to secondary pollutants: The above definition of the human toxicity category implies that only primary pollutants are considered (‚direct emission'). In line with that, the secondary photo-oxidants are subsumed under a different impact category, with the possibility of using human toxicity category indicators for photo-oxidants being mentioned. For the abovementioned pragmatic reason of coherent modeling, the health effects of secondary sulfate and nitrate particles will be included in the human toxicity category here rather than within the category of acidification, where these aerosols also play an important role.

General Framework for the calculation of category indicators: Category indicators for human toxicity are generally calculated by multiplying the emitted mass M of a pollutant p with a fate and exposure factor F and an effect factor E. In the general case where the transfer of the pollutant across different environmental media or compartments (e.g. air, water, soil) needs to be considered, the category indicator ΔD_p^{nm} (D for damage) characterizing the effect in the compartment m of pollutant p emitted into the initial compartment n is written as

$$\Delta D_p{}^{nm} = E_p{}^m \, F_p{}^{nm} \, M_p{}^n \qquad (2.3)$$

with

$M_p{}^n$ mass of pollutant p being emitted into the initial medium n (air, water or soil)

$F_p{}^{nm}$ Fate and exposure factor for substance p emitted into the initial medium n and transferred into medium m, taking into account the propagation, degradation, deposition, intermedia transfer and food chain / bioconcentration routes

$E_p{}^m$ Effect factor representing the degree of toxicity of substance p in medium m (air, water, soil or food chain) (Jolliet et al. 1996:53).

$\Delta D_p{}^{nm}/M$ is called damage factor (Hofstetter 1998:102). In the present work, only toxic pollutants emitted into air and affecting human health by inhalation are of interest. Multimedia transfer therefore will not be considered, and the index n specifying the initial compartment can be omitted. Instead of indicating receiving media other than air, the index m can be used in this·case to distinguish between the primary emitted pollutant (m = 1) and secondary pollutants (m > 1) that may be formed. The fate and exposure factors for the secondary pollutants include information about their rate of formation from the primary pollutant.

The above general definitions of fate and effect factors do not explicitly consider the number and spatial distribution of the receptors affected by the pollutants, e.g. the population density in the case of air pollutants. In order to be applicable to the general case of an inhomogeneous population density distribution around the emission source, an *effective* population density $\rho_{eff\,p}{}^m$ will be added here to characterize that distribution in relation to the spatial distribution of the primary pollutant p (m = 1) or its associated secondary pollutants (m > 1)[9]. With these adaptations suitable for the case of air pollutants, equation (2.3) reads

$$\Delta D_p{}^m = E^m \, \rho_{eff\,p}{}^m \, F_p{}^m \, M_p \qquad (2.4)$$

Physical interpretation of fate factors, effect factors and effective population densities: With regards to the dimensions of E and F (i.e. the type of units in which they are expressed), a number of different possibilities and combinations exists, depending on the type of toxicity (acute versus chronic) and exposure (short peaks versus continuous), the method used to determine the toxicity (toxicological versus epidemiological) and the completeness of modeling the fate of the pollutant in the environment (proxy indicators versus complete modeling).

Effect factors are generally expressed as the number or duration of health-related incidences divided by an independent effect variable. In the case of the epidemiological effect factors used in the present work, the independent effect variable has the dimension of (persons × exposure time × concentration). This product of the number of exposed persons, the pollutant concentration to which they are exposed and the time duration of the exposure will be called *population exposure* in the following and denoted with the symbol PE. It will be given in units of

---

[9] Hofstetter (1998:258, 327) instead includes the effective population density in the effect factor E.

1 person×μg/m^3×a, which corresponds to 1 person being exposed to a constant pollutant concentration of 1 μg/m^3 over 1 year (1 a). In the same way, an *area exposure* can be defined as the product of exposed surface area, pollutant concentration and time duration of the exposure.

According to Jolliet et al. (1996:55), a fate and exposure analysis is complete when it relates the mass of the emitted pollutant to the independent effect variable. In the case of air pollutants characterized by the above effect factors, the product of fate factor and effective population density then has the dimension of population exposure per mass of emitted pollutant and is expressed in units of 1 person×a/m^3. The fate factor accordingly represents the area exposure per mass of emitted pollutant. It has the dimension of area×time/volume and is typically expressed in units of 1 m^2×a/m^3 (= 1a/m).

The concrete meaning of the fate factor, the effect factor and the effective population density will in the following be illustrated further for the case of air pollutants by relating them to physical parameters of emission situations. For that purpose, consider an individual person at a location $\underline{x}$ (two-dimensional position vector) on the surface of the earth who is exposed to a concentration $c(\underline{x},t)$ of a particular pollutant varying over time t. Within the infinitesimal time interval dt, this exposure causes a damage of size $d(c(\underline{x},t))$ dt, where the function $d(c)$ is the *individual exposure response function*. In the special case of $d(c)$ being proportional to c, it is called *linear*, but in general it will be a nonlinear function with $d(0) = 0$. The total damage D to the individual is given by the time-integral

$$D(\underline{x}) = \int d(c(\underline{x},t))\, dt \qquad (2.5)$$

When an incremental emission of the pollutant located at $\underline{x}_0$ is considered, it is suitable to write the total pollutant concentration as the sum

$$c(\underline{x},t) = b(\underline{x}) + \Delta c(\underline{x},t) \qquad (2.6)$$

of a time-independent background concentration $b(\underline{x})$ due to all other emission sources and the *incremental* concentration $\Delta c(\underline{x},t)$ caused by the emission process[10]. The incremental damage $\Delta D(\underline{x})$ due to the additional emission source is the difference between the damages with and without the presence of $\Delta c(\underline{x},t)$, i.e.

$$\Delta D(\underline{x}) = \int d(b(\underline{x})+\Delta c(\underline{x},t)) - d(b(\underline{x}))\, dt =$$
$$\int d'(b(\underline{x}))\, \Delta c(\underline{x},t) + O(\Delta c^2\,(\underline{x},t))\, dt \qquad (2.7)$$

where a formal Taylor expansion was used in the second integral. $d'(c)$ thereby denotes the first derivative of $d(c)$ and $O(\Delta c^2\,(\underline{x},t))$ is used as a shorthand for the terms of second or higher order in Δc. It should be noted that these higher-order terms are identical to zero for *linear* exposure response functions $d(c)$. Furthermore, they can be neglected in good approximation if $\Delta c(\underline{x},t) \ll b(\underline{x})$, i.e. if the considered emission source contributes only *marginally* to the overall immissions. If the higher-order terms can be neglected for either one of these two reasons, the incremental damage becomes

---

[10] The background concentration b may also be a function of time, but its variation typically occurs over much longer timescales than the duration of the considered individual processes and is therefore neglected here.

$$\Delta D(\underline{x}) = d'(b(\underline{x})) \int \Delta c(\underline{x},t) \, dt \tag{2.8}$$

The time integral in (2.8) is actually independent of the time characteristics of the additional emission process (i.e. the time-dependence of its emission rate $Q(\underline{x_0},t)$) and only depends on the total emitted mass

$$M = \int Q(\underline{x_0},t) \, dt \tag{2.9}$$

of the pollutant. This is due to the linearity of the partial differential equations relating $Q(\underline{x_0},t)$ to $\Delta c(\underline{x},t)$ and the associated possibility of a linear superposition of its solutions[11]. Therefore, the incremental damage due to an emission process $Q(\underline{x_0},t)$ is independent of its time characteristics and only depends on the total emitted mass M provided that the higher order terms in (2.7) can be neglected, which holds for linear exposure response functions or marginal emission processes in the sense defined above.

If the linearization of (2.7) is not possible, the time characteristics of the emission may play a role by way of determining which ranges of the exposure response function d(c) are relevant. To provide an extreme example, if d(c) exhibits a distinct threshold above the background concentration (i.e. no damages occur without the incremental emission), the time characteristics of the emission may, in addition to the emitted mass, determine whether the threshold will be exceeded or not.

In the following, only the cases of linear exposure response functions or marginal contributions will be considered further, since these turn out to be the cases of practical relevance for the case study presented in chapters 4 and 5. In this case, it is suitable to define the *incremental individual exposure* $\Delta PE(\underline{x})$ as

$$\Delta PE(\underline{x}) \equiv \int \Delta c(\underline{x},t) \, dt \tag{2.10}$$

Due to their independence of the time characteristics of the emission, the incremental individual exposure (2.10) and the resulting incremental individual damage (2.8) can be calculated assuming an arbitrary constant emission rate $Q = M/T$, where T is the arbitrary duration of the emission. For the same reason, instead of calculating a time-dependent concentration $\Delta c(\underline{x},t)$ for an emission pulse of duration T and evaluating the infinite time integrals in (2.8) and (2.10), these equations can also be evaluated by using the steady-state concentration $\Delta c(\underline{x})$ (which is proportional to Q) and limiting the time-integrals to an exposure time of duration T (Heijungs 1995). Equation (2.8) then reads

$$\Delta D(\underline{x}) = M/Q \; d'(b(\underline{x})) \; \Delta c(\underline{x}) \tag{2.11}$$

where the quotient $\Delta c(\underline{x})/Q$ is independent of Q. The incremental damage according to equation (2.11) refers to one exposed person. In order to obtain the incremental damage ΔD occurring in the entire population, it needs to multiplied with the number of people $\rho(\underline{x}) \, dA$ in the area element dA around $\underline{x}$ and integrated over the whole area A surrounding the source, i.e.

$$\Delta D = M/Q \; \iint d'(b(\underline{x})) \; \Delta c(\underline{x}) \; \rho(\underline{x}) \, dA \tag{2.12}$$

---

[11] The differential equations are linear as long as no non-linear chemical reaction schemes such as for secondary pollutants like ozone are involved. Otherwise, the marginality assumption needs to be invoked in order to linearize the nonlinear equations.

By expanding with the product of the three area integrals of the individual factors in the integrand, this can be written in the form of equation (2.4) with (indices omitted)

$$F = 1/Q \iint \Delta c(\underline{x}) \, dA \tag{2.13}$$

$$E = 1/A \iint d'(b(\underline{x})) \, dA \tag{2.14}$$

$$\rho_{eff} = \iint d'(b(x)) \, \Delta c(\underline{x}) \, \rho(\underline{x}) \, dA \, / \, (\, 1/A \iint d'(b(x)) \, dA \times \iint \Delta c(\underline{x}) \, dA \,) \tag{2.15}$$

This shows that the effect factor E is equal to the average slope of the exposure response function. The effective population density is the weighted average of the population density distribution around the source with $d'(b(x))$ and $\Delta c(\underline{x})$ being the weighting factors. In the case of a spatially independent exposure response slope $d'(b(\underline{x})) = E = $ const, which may either be due to a constant background concentration $b(\underline{x}) = b$ or due to a linear exposure response function (the latter case being of particular interest here), equations (2.12) and (2.15) reduce to

$$\Delta D = E \ M/Q \iint \Delta c(\underline{x}) \, \rho(\underline{x}) \, dA \tag{2.16}$$

$$\rho_{eff} = \iint \Delta c(\underline{x}) \, \rho(\underline{x}) \, dA \, / \iint \Delta c(\underline{x}) \, dA \tag{2.17}$$

In this case, it is useful to define the *incremental population exposure* (in analogy to the incremental individual exposure) as

$$\Delta PE = M/Q \iint \Delta c(\underline{x}) \, \rho(\underline{x}) \, dA \ = \Delta D/E = M \, F \, \rho_{eff} \tag{2.18}$$

It should be emphasized again that these expressions of the variables ΔD, E, F and ρ_{eff} formally defined in equation (2.4) in terms of physical parameters only hold when either the incremental emission contributes marginally to the overall pollutant concentrations, or when the exposure response function d(c) is linear. In particular, this limitation applies to the circumstance that the incremental damage only depends on the emitted mass M of the pollutant, as it is suggested by equation (2.4), and not on the time characteristics of the emission.

In the following, existing methods for the determination of fate and exposure factors, effective population densities and effect factors will be discussed. A comprehensive review is beyond the scope of this chapter. Rather, the focus is on methods pertaining to air pollutants with carcinogenic or respiratory health effects, since these are of relevance in the present work.

Effect Factors: Effect factors can be based either on toxicological or on epidemiological data. Toxicological data are typically derived from tests on animals which are exposed to a single chemical under controlled conditions. The practical limits of available time and number of animals for a toxicological study require exposures to be much larger than those found in the environment. For the purpose of assessing health effects of environmental pollution on humans, the results of toxicological studies therefore need to be extrapolated from animals to humans and from high to low doses or exposures.

Epidemiological studies, on the other hand, examine actually occurring health effects in defined groups of the human population and try to relate the observed effects to the exposure situation of the investigated group of persons. While extrapolations as in the case of toxicological studies are therefore not required, problems of the epidemiological approach are associated with the fact that the exposure situation in the past cannot always be precisely reconstructed. Furthermore, the examined groups are typically exposed to more than one agent causing the effect of interest, such that confounding factors need to be taken care of.

In the ideal case, toxicological or epidemiological studies would provide information about the dependence of the effects ('response') on the external exposure or the absorbed dose of a pollutant, with exposure being the typically used independent variable in the case of air pollutants. The effect factor E would then be determined as the slope of the exposure-response function. In general, the effect factor applying to a particular marginal increase of the exposure (associated with the functional unit) would be a function of the background concentration. In the absence of more detailed data, however, the exposure-response function is often assumed to be homogeneously linear and therefore represented by a constant effect factor. This linearity assumption may be justified or commonly accepted to varying degrees, as will be discussed below for the examples of air pollutants with carcinogenic and respiratory health effects which are of particular interest here.

In the absence of quantitative information about exposure-response slopes, a common assumption within the toxicity characterization in LCA is to assume that they are inversely proportional to some effect thresholds such as No Observed Adverse Effect Levels (NOAEL) from toxicological studies or legal limit values. On the basis of such effect factors, which do not contain information about the absolute size of the effects, the human toxicity of a substance is expressed relative to a reference substance. Fate analyses used in connection with these relative effect factors may be complete, but are not necessarily so.

Recently, due to progresses in fate modeling and the availability of exposure-response slopes, absolute effect factors have become available at least for some groups of toxic substances. Many contributions to the development of absolute effect factors come from studies that are concerned with external costs, e.g. of energy or transport systems (European Commission 1995, 1998). While different types of damages to human health are aggregated on the basis of monetary units within externality studies, the monetarization is only the last of many steps of the impact pathway analysis and can also be omitted in order to derive absolute effect factors suitable for LCA (Spadaro and Rabl 1999).

A classification of human health effects based on their mechanisms (Boguski et al. 1996:2.30) distinguishes

- carcinogenesis
- irritant, corrosive effects
- respiratory effects
- central nervous system effects
- allergenicity, sensitisation
- methemoglobinemia, blood diseases
- effects of odours
- cardiovascular system effects
- behavioural effects
- bone or renal effects.

In the following, only air pollutants acting as carcinogens or causing respiratory effects will be considered. A number of quantitative studies show that these two effect categories are responsible for a large fraction of the actually occurring health damages due to environmental pollution (Hofstetter 1998:108). In particular, the airborne traffic emissions of interest for the case study presented in chapters 4 and 5 are associated with health effects mainly in these two categories.

Based largely on the results of epidemiological studies, Hofstetter (1998) has derived effect factors suitable for the use in Life Cycle Impact Assessment for a large number of substances falling into these two categories. Only a few main points regarding the derivation of these effect factors are highlighted in the following, and the numerical values of the factors will be reproduced here only for those substances considered in the case study. These are Diesel soot particles, benzene, benzo[a]pyrene, formaldehyde, acetaldehyde and 1,3-butadiene as carcinogens and particulate matter (including Diesel soot particles), sulfur dioxide (SO_2), nitrogen oxide (NO_x), secondary sulfate and nitrate aerosols and ozone (O_3) formed from the precursors NO_x and volatile organic compounds (VOC) as pollutants leading to respiratory health effects.

Carcinogenic Effects: Evidence for the human carcinogenity of single substances usually comes from epidemiological studies of people exposed to high concentrations of these substances at the workplace, sometimes in combination with results from animal tests. Epidemiological studies among the general population are generally not useful since the additional effects to be expected from the exposure to the ambient concentrations of the substance of interest are too small compared to the overall occurrence of cancers, which is dominated by lifestyle factors such as smoking. Even occupational epidemiological studies, like epidemiological studies in general, have to deal with the problems of multi-agent exposure and incomplete knowledge about past exposures. The latter problem is particularly severe in the case of carcinogenic substances due to the long latency times. According to the degree of available evidence, carcinogenic agents can be classified qualitatively. One widely used classification scheme is that of the International Agency for Research on Cancer (IARC), which is reproduced in table 2.4 (Hofstetter 1998:233-238).

In addition to the qualitative information about the carcinogenity of an individual substance, quantitative relations between exposure or dose and response can sometimes be determined from occupational epidemiological studies or from animal bioassays. In order to quantify the cancer risks of individual substances at ambient concentrations, an extrapolation to lower concentrations is necessary. A number of assumptions are required for that purpose, which are often the subject of considerable debate. However, the following assumptions are currently being used and accepted in the field of health policy:

- The development of cancers is a function of the time integrated exposure, i.e. the time pattern of the experienced concentrations does not play a role.
- For genotoxic chemical agents (i.e. those which initiate or promote the development of cancers by causing mutations), there is no threshold for the effect, because they damage the DNA in a stochastic and irreversible way.

Table 2.4. Classification scheme for carcinogenic substances with regards to the degree of available evidence used by the International Agency for Research on Cancer, and number of agents classified so far (768 in total) (Hofstetter 1998:233-234)

group	number of agents	description
1	60	The agent, mixture, or exposure circumstance is carcinogenic to humans.
2A	51	The agent, mixture, or exposure circumstance is probably carcinogenic to humans (a positive association has been observed between exposure and human cancer for which a causal interpretation is credible, but chance, bias, or confounding could not be ruled out with reasonable confidence; there is also sufficient evidence of carcinogenity in experimental animals).
2B	206	The agent, mixture, or exposure circumstance is possibly carcinogenic to humans (there is sufficient evidence of carcinogenity in experimental animals, but no adequate data on cancer in exposed humans).
3	450	The agent, mixture, or exposure circumstance is not classifiable as to its carcinogenity to humans (this Group applies if no other category is used).
4	1	The agent, mixture, or exposure circumstance is probably not carcinogenic to humans.

- The measured dose-response or exposure-response curves are extrapolated linearly (i.e. along a straight line) towards the origin, which may lead to an overestimation of the effects (Hofstetter 1998:238).

Based on these assumptions, the exposure-response function for a carcinogenic agent can simply be characterized by its constant slope, which is commonly expressed in terms of the *unit risk factor* (also denoted as unit lifetime risk or unit risk). In the case of inhalation, the unit risk is an "estimate of the probability that an average individual will develop cancer when exposed to a pollutant at an ambient concentration of one microgram per cubic meter ($\mu g/m^3$) for the individuals life (70 years)" (Hofstetter 1998:237). Unit risks are given in $(\mu g/m^3)^{-1}$, but their reference to the exposure time of 70 years should be kept in mind.

Unit risks are not available for all of the 317 substances in the IARC classes 1,2A and 2B. The largest database of unit risk factors is maintained and continuously updated by the U.S. Environmental Protection Agency (EPA 1996, Integrated Risk Information System IRIS). For a smaller number of substances (10), more than one estimate of the unit risk is available. Among these substances are three which are of interest here, namely benzene, benzo[a]pyrene and Diesel soot particles. The unit risk estimates for the former two vary by a factor between 2 and 3, whereas the estimates for Diesel soot particles vary by more than a factor ten because they are based on very different size ranges and chemical compositions. Table 2.5 lists the available unit risks for those substances of relevance for the case

Table 2.5. Unit risks due to inhalation for selected carcinogenic substances relevant for the case study in chapters 4 and 5. Where more than one estimate is available, the most recent one is used (Hofstetter 1998:240, 428). The other estimates will be considered for sensitivity analyses where necessary.

substance	IARC group	unit risk $[(\mu g/m^3)^{-1}]$				
		used here	WHO 1996a	WHO 1987	EPA 1996	LAI 1992
benzene	1	6 E-6	6 E-6	4 E-6	8,3 E-6	9 E-6
1,3-butadiene	2A	2,8 E-4			2,8 E-4	
PAH as benzo[a]pyrene[a]	2A	8,7 E-2	8,7 E-2	9 E-2	7 E-2	7 E-2
Diesel soot particles	2A	3,4 E-5	3,4 E-5 [b]		2-10 E-5	7 E-5
formaldehyde[a]	2A	1,3 E-5			1,3 E-5	
acetaldehyde	2B	2,2 E-6			2,2 E-6	

IARC International Agency for Research on Cancer, *PAH* Polycyclic Aromatic Hydrocarbons
[a] For formaldehyde and benzo[a]pyrene, 63 % and 96 %, respectively, of the total damage is due to the uptake of food and drinking water (Hofstetter 1998:430).
[b] WHO 1996b.

study presented in chapters 4 and 5. In line with Hofstetter (1998:240), the latest estimate will be used, if more than one value is available. The other ones will be considered for sensitivity analyses where necessary.

In order to provide an approximate quantitative measure of the uncertainty associated with the unit risk factors as well as other effect factors and fate factors, it is convenient to assume lognormal distributions, i.e. distributions which are normal on a logarithmic scale. For a lognormal distribution characterized by a geometric mean μ_g and a geometric standard deviation σ_g, the 68 % confidence interval is $[\mu_g/\sigma_g, \mu_g \times \sigma_g]$ and the 95 % confidence interval is $[\mu_g/\sigma_g^2, \mu_g \times \sigma_g^2]$ (European Commission 1998:87-89; Hofstetter 1998:424-426). The uncertainties associated with the unit risks were estimated to be $\sigma_g^2 = 2, 3, 5$ and 10 for carcinogens in the IARC groups 1, 2A, 2B and 3, respectively by Yetergil Kiefer (1997, quoted after Hofstetter 1998:246).

To keep track of dimensions in later applications, it is suitable to transform the unit risks into an effect factor indicating cancer incidences per population exposure. This is easily done by observing that the population exposure corresponding to the unit risk factor is 1 person $\times$ 70 years $\times$ 1 ($\mu g/m^3$), and that multiplying the risk (in the sense of a probability or relative frequency) by a number of persons yields a number of cancer incidences.

Furthermore, it is useful to transform the variable ‚cancer incidences' into measures that facilitate a comparison with other effects on human health. Such measures relate to the general categories of mortality and morbidity. With regards to mortality, it is becoming increasingly common to use the Years of Life Lost (YLL) as a measure instead of the number of fatalities, since this allows to distinguish between fatalities occurring at different ages. The Years of Life Lost are basically the difference between the age at which the fatality occurs and the age the individual would have reached otherwise. Of course, the latter can only be determined in a hypothetical way based on mortality statistics. Different methodological options in this regard are described by Hofstetter (1998:175-176), but shall not be discussed here in detail. Furthermore, a discount rate to account for time preferences and a weighting of ages (with a maximum for persons in their 20s) are sometimes used in the calculation of the YLL, but these options shall not be considered here.

Besides the probable fatal outcome of a cancer incidence (the death rate after ten years being 74 % on average over all types of cancer, Hofstetter 1998:252), the preceding morbidity effects also need to be considered. Similar to the calculation of Years of Life Lost, the overall duration of each type of non-fatal effect may be taken as a measure. Due to the large number of different types of morbidity effects in general, a further aggregation is useful. This requires the introduction of weights representing the relative severity of the various types of effects. Such a weighting procedure may also include fatal effects in order to provide for an overall aggregation of both fatal and non-fatal effects.

The weighting of different types of health outcomes clearly represents a normative presupposition. It is therefore not surprising that various weighting sets and associated health indices have been proposed. None of them has so far gained widespread acceptance, but among the well-known ones are the Quality Adjusted Life Years (QALY, Baldwin et al. 1990) and the Disability Adjusted Life Years (DALY, Murray 1996). Furthermore, studies determining external costs of technical systems (e.g. electricity generation) provide aggregations of various health effects in monetary terms (e.g. European Commission 1995, 1998). A review of the various approaches is given by Hofstetter (1998:164-171).

The DALY indicator is defined as the sum of Years of Life Lost (YLL) and Years Lived Disabled (YLD). In the absence of age-weighting and discounting of the future, the latter are given as the product of the duration L of a disability and a respective disability weight DW, which assumes a value between 0 and 1. Accordingly, the DALY indicator for a given type of disease m has the structure

$$\text{DALY}_m = \text{YLL}_m + \text{YLD}_m = \text{YLL}_m + \text{DW}_m \, \text{L}_m \qquad (2.19)$$

The crucial determination of the disability weights DW_m for a number of diseases was carried out by an international panel of health care providers at a meeting convened by the World Health Organization (WHO) in Geneva in August 1995. The results of that meeting agreed well with the average results of nine prior test meetings.

On this basis, Hofstetter (1998:249-256) has calculated YLL, YLD and DALY values for a large number of carcinogenic substances. For the selection of substances of relevance here, these are shown in table 2.6. They are normalized to the population exposure, which is convenient for later use. For this selection of substances, the number of Years of Life Lost due to one cancer incidence varies between about 12 and 16, whereas the Years of Life Disabled (YLD) amounts to between 0,3 and 0,5 years. The DALY values for the considered cancer incidences are therefore dominated by the YLL values. The geometric standard deviations of the DALY values listed in table 2.6 combine the estimated uncertainties of the unit risks and of the YLD and YLL values per cancer incidence.

Table 2.6. Effect factors due to inhalation for the selection of carcinogenic substances relevant for the case study presented in chapters 4 and 5 (Hofstetter 1998:240, 255, 428)

substance	IARC group	unit risk $[(\mu g/m^3)^{-1}]$	CI / PE [1/(persons a $(\mu g/m^3)$)]	YLL / PE	YLD / PE	DALY / PE	σ_g^2
				[1/(persons $(\mu g/m^3)$)]			
benzene	1	6 E-6	8,6 E-8	1,4 E-6	3,9 E-8	1,4 E-6	4,6
1,3-butadiene	2A	2,8 E-4	4,0 E-6	5,0 E-5	1,9 E-6	5,2 E-5	5,7
PAH as benzo[a]pyrene[a]	2A	8,7 E-2	1,2 E-3	2,0 E-2	3,6 E-4	2,0 E-2	7,0
Diesel soot particles	2A	3,4 E-5	4,9 E-7	7,7 E-6	1,4 E-7	7,8 E-6	5,7
formaldehyde[a]	2A	1,3 E-5	1,9 E-7	2,2 E-6	7,6 E-8	2,3 E-6	7,6
acetaldehyde	2B	2,2 E-6	3,1 E-8	4,0 E-7	1,5 E-8	4,1 E-7	8,2

CI cancer incidences, *PE* population exposure, *YLL* Years of Life Lost, *YLD* Years Lived Disabled, *DALY* Disability Adjusted Life Years (the latter three without age weighting and without discounting), σ_g geometric standard deviation for DALY/PE, *IARC* International Agency for Research on Cancer
[a] For formaldehyde and benzo[a]pyrene, 63 % and 96 %, respectively, of the total damage is due to the uptake of food and drinking water (Hofstetter 1998:430).

Respiratory health effects: The second group of substances of interest for the case study on natural gas vehicles are air pollutants causing respiratory health effects. The following substances are considered: particulate matter (PM) of various size ranges, nitrogen dioxide (NO_2), sulfur dioxide (SO_2) and the associated secondary nitrate and sulfate aerosols, carbon monoxide (CO) and tropospheric ozone (O_3) formed from nitrogen oxides and volatile organic compounds (VOC) as precursors.

Effect factors for these pollutants are provided in a review of a large number of epidemiological studies (Pilkington et al. 1997) that was conducted within the ExternE project on external costs of air pollution by electricity generation and transport (European Commission 1995, 1998; Friedrich et al. 1998). The health endpoints for which exposure response functions are available from this review comprise chronic mortality (i.e. mortality occurring a long time after the exposure), acute mortality (i.e. mortality occurring within a few days after an exposure, mostly affecting people already weakened by other factors) and a large number of non-fatal respiratory health effects.

One important property of the exposure-response functions reviewed by Pilkington et al (1997) is that no threshold can be identified at the population level below which no effects do occur. While thresholds may exist for each person individually, the variation of the individual threshold levels within the population is such that particularly susceptible persons (e.g. elderly or children) are affected by low pollutant concentrations such as in rural areas as well. Both Hofstetter (1998:291) and the ExternE studies (European Commission 1998, Friedrich et al. 1998) therefore base their assessment of the health effects of the above pollutants on the assumption that no threshold exists. Furthermore, both use, for the time being, as a reasonable approximation the assumption that the exposure-response functions are linear. Under this prerequisite, they can be characterized by their constant slope, which is expressed as cases per population exposure, typically in units of cases / (person×year×µg/m^3).

The large number of effect factors (59) for the variety of effects associated with the pollutants considered here shall not be listed in detail (see Hofstetter 1998:321-322; European Commission 1998), since it turns out that the aggregated effects in terms of Years Lived Disabled (YLD) and Years of Life Lost (YLL) are dominated by only two or three effects for each pollutant. This refers to the YLD, YLL and DALY values calculated by Hofstetter (1998:329-335) on the basis of the effect factors provided by Pilkington et al. (1997) (table 2.7). In their derivation, a value of 0,75 YLL per incidence of acute mortality (with an uncertainty of $\sigma_g^2 = 5$) and 6,6 YLL per incidence of chronic mortality (with an uncertainty of $\sigma_g^2 = 3$) were assumed as additional input data. Furthermore, disability weights for the non-fatal effects were estimated by analogy for those effects not covered in the original definition of the DALY concept (Murray et al. 1996). With one exception, this led to disability weights of either 0,05 or 0,1. In this context, it should be noted that the dominant contribution of chronic bronchitis to the overall non-fatal effects of primary particulate matter and secondary sulfate and nitrate aerosols is due to its long duration of 40 years (for adults), compared to typical durations in the order 1 to 3 days for the other health endpoints.

Similar to the IARC classification for carcinogenic substances, the effect factors listed in table 2.7 can be differentiated with regards to the quality of the evidence supporting a causal interpretation of the observed epidemiological associations. Quality criteria in this regard are the biological plausibility of the existence of an effect mechanism and the number and consistency of the respective high-quality epidemiological studies. In this regard, the quality of the evidence for the health effects due to ozone is high among the pollutants causing respiratory health effects, but nevertheless lower than in the case of the carcinogens discussed above. In quantitative terms, the effect factors for ozone can be assigned an estimated uncertainty of $\sigma_g^2 = 15$. This also applies to primary particles (PM 10 or PM 2,5) and secondary sulfate aerosols. Many studies associate these particles with respiratory health effects, and three different effect mechanisms are being discussed. Only one of these mechanisms would be applicable to nitrate aerosols, such that the evidence is of lower quality for nitrate aerosols than for the other particles ($\sigma_g^2 = 34$). In the case of SO_2 and NO_2, it is not clear to what extent they only act as surrogate indicators for the presence of other pollutants (sulfates in the case of SO_2, many other traffic related pollutants in the case of NO_2). While limited evidence is available for an independent link of SO_2 to health effects ($\sigma_g^2 = 25$), this does not apply to NO_2 ($\sigma_g^2 = 51$). Finally, the evidence for the health effects of CO is of low quality ($\sigma_g^2 = 51$), since only a few studies show a link to health effects at ambient concentrations and the toxicological mechanisms are only known to be effective at much higher concentrations (Hofstetter 1998:324-335).

Table 2.7 shows that the effect factors for the primary pollutants SO_2 and NO_x are more than two orders of magnitude smaller than those of the associated secondary sulfate and nitrate aerosols. In combination with the population exposures per emitted mass listed in table A.2 in the appendix, the impact of SO_2 (NO_x) will therefore be at most 10 % (5 %) of the impact of the secondary sulfates (nitrates).

Table 2.7. Effect factors for respiratory health effects of selected airborne pollutants relevant for the case study in chapters 4 and 5 (Hofstetter 1998:335, 440). In addition to the total effect factors for each substance, the dominant contributions to YLD and YLL are listed.

pollutant health endpoint	YLD / PE	YLL / PE [1 / (persons $\times$ $\mu g/m^3$)]	DALY / PE	σ_g^2
PM 10 / nitrates				
chronic bronchitis	7,8 E-5		7,8 E-5	
chronic mortality		2,1 E-4	2,1 E-4	
total	1,0 E-4	2,1 E-4	3,1 E-4	PM 10: 15 nitrates: 34
PM 2,5 / sulfates				
chronic bronchitis	1,2 E-4		1,2 E-4	
chronic mortality		3,4 E-4	3,4 E-4	
total	1,7 E-4	3,4 E-4	5,1 E-4	15
ozone				
asthma attacks	1,2 E-6		1,2 E-6	
symptom days	4,6 E-6		4,6 E-6	
acute mortality		3,7 E-6	3,7 E-6	
total	7,3 E-6	3,7 E-6	1,1 E-5	15
SO$_2$				
total (acute mort.)		4,5 E-6	4,5 E-6	25
NO$_x$ as NO$_2$				
total (acute mort.)		2,1 E-6	2,1 E-6	51
CO				
total (acute mort.)		9,1 E-8	9,1 E-8	51

PE population exposure, *YLD* Years Lived Disabled, *YLL* Years of Life Lost, *DALY* Disability Adjusted Life Years, σ_g geometric standard deviation, *mort.* mortality

Furthermore, the quality of the underlying evidence is also lower for the primary pollutants in each case. Therefore, the direct health effects of SO$_2$ and NO$_x$ will be neglected in the case study in chapters 4 and 5. The same applies to the health effects of carbon monoxide (CO) because of the low quality of the evidence. A comparison of the effect factors for the carcinogenic effects and the respiratory health effects of Diesel particles (tables 2.6 and 2.7, with Diesel particles being subsumed under PM 2,5 in the latter case) shows that both the YLL value for the respiratory chronic mortality as well as the YLD value are considerably higher than the YLL factor for carcinogenity.

Fate factors and effective population densities: In order to determine fate factors and effective population densities in the sense of a full fate and exposure analysis (Jolliet et al. 1996:55), the distribution of the pollutants coming from a specified emission source within the environment need to be determined in relation to the distribution of the receptors (human beings). Various types of models can be used to determine the distribution of pollutants in the environment.

Since the method and the case study presented here (chapters 3-5) are only concerned with air pollutants and inhalation as an exposure pathway[12], the use of multimedia models is not required. Furthermore, due to their presently still low spatial resolution, such models would not be suitable to deal with the question of the spatial differentiation of impacts. Multimedia models will therefore not be considered here. Introductions to the principles of multimedia modeling are given by Mackay (1991), Seigneur (1993) and Klöpffer (1996). A review of some frequently used multimedia models can be found in (Cowan et al. 1995). Multimedia fate factors to be used in LCA were calculated for a large number of substances, e.g., by Guinée and Heijungs (1993) and by Hofstetter (1998).

Air pollution models, which consider the distribution $\Delta c(\underline{x})$ of pollutants in the atmosphere around specified emission sources, exist in a large variety for different purposes of application. They can be distinguished according to their spatial scale and resolution (ranging from street canyons to the entire globe), the amount of meteorological input data they require (ranging from a few simple parameters to the solution of complex windfield models), and the degree to which they consider chemical transformations (from no chemistry for inert primary pollutants to complex atmospheric chemistry schemes, e.g. for photo-oxidant formation). Accordingly, the resources required for their application range from a sheet of paper to insert numerical values into a few analytical formulae to considerable computing times on powerful scientific computers.

In the context of LCA, the use of complex models with large amounts of required input data is prohibitive since the data collection for the emission inventory is already resource intensive by itself. This limits the spectrum of possibilities to either using rather simple models with only a few input parameters or using the *results* of more complex models, provided that these are easily available (in tables or databases) for practically *all* emission situations that may have to be considered in an LCA.

An overview of air pollution modeling is beyond the scope of this chapter. Reviews are provided by van Dop (1985), Zanetti (1990), Moussiopoulos et al. (1996), Zenger (1998) and Seinfeld and Pandis (1998). In the following, only those models that are used in chapter 3 or have been used in other methods for Life Cycle Impact Assessment are briefly discussed.

A basic distinction concerns the role of empirical observations of pollutant concentrations. *Empirical* air pollution models are based on establishing statistical correlations between observed spatial distributions of pollutant concentrations and the locations where the pollutants are emitted. In the context of LCA, Jolliet (1994) and Jolliet and Crettaz (1997) followed the empirical approach to determine fate factors for 20 pollutants for which measurements from a large number of stations

---

[12] For the considered air pollutants, it is known that only the inhalation pathway is relevant with regards to impacts on human health. This holds with the exception of formaldehyde and benzo[a]pyrene. The relevance of other exposure pathways for these pollutants (mainly food intake) for the results of the case study on natural gas and Diesel vehicles will be considered in chapter 5.

within the reference area as well as the emitted amounts (natural and man-made) are available (e.g. PM, SO_2, NO_x, CO). The fate factors are determined by relating the average pollutant concentration to the total emission flow from *all* sources in the reference area. For 100 additional pollutants, fate factors were determined as a function of their atmospheric residence times by interpolating between the empirically determined values. The considered reference areas were Switzerland and the whole world, respectively.

Limitations of this approach with regards to air pollutants concern the representativity of the locations where the pollutant concentrations are measured and the necessity to budget the import and export of pollutants in the case of reference areas other than whole world (Hofstetter 1998:220). The main limitation of interest here, however, is that the method does not consider individual emission sources. It therefore does not allow for a differentiation of fate factors and effective population densities according to the location and height of an emission.

In order to introduce such differentiations, concentrations related to individual emission sources need to be considered rather than average concentrations resulting from a large number of emissions. Empirical models can be used to consider an individual source as well, but only within the range where that source dominates the overall measured concentration. Measurements around hiways in Germany have shown that the pollutant concentrations (e.g. of CO) have decreased down to the background level at a few hundred meters away from the hiway (Koch and Windt 1988; Kuhler et al. 1994). However, as will be shown in chapter 3, much larger distances (up to some hundred or thousand kilometers) need to be considered to cover the total damage caused by an individual emission source. In other words, most of the damage typically does not occur in the area where the emission source dominates the overall concentration, but in the much larger area where the individual source is only one of many contributors to the background concentration.

Empirical models can therefore not be used to determine the total impact caused by individual emission sources. Instead, mathematical models need to be used that theoretically simulate the atmospheric processes affecting the distribution of pollutants around an emission source. Among the large variety of existing models, only the two types of models that will be used in the present work (chapter 3), namely Gaussian plume models and Lagrangian trajectory models, are briefly described in the following.

Gaussian plume models are relatively simple atmospheric dispersion models which are applicable to primary pollutants. Despite their theoretical limitations, they are widely used in practice, e.g. for regulatory purposes. In Germany, for example, a variety of regulations such as the Technical Guidance Document Air (Technische Anleitung Luft) pertaining to the Air Pollution Control Act (Bundes-Immissionsschutzgesetz) and several guidelines (Richtlinien) of the German engineering society VDI (Verein Deutscher Ingenieure) are based on Gaussian dispersion models.

Gaussian plume models are based on an analytical solution of the time-independent partial differential equation describing the atmospheric advection and diffusion of pollutants (Seinfeld and Pandis 1998:942, eq. 18.65) under the prerequisites that wind direction and wind speed u as well as the parameters describing turbulent atmospheric diffusion (within the so-called K-theory) are constant in space and time, and that the turbulent diffusion in the wind direction is small compared to the advective transport, which holds for wind speeds larger than about 1m/s. In the stationary case, the incremental pollutant concentration around a continuous emission source with emission rate Q at an effective emission height[13] h above the ground located in flat terrain with no obstacles, with the wind blowing towards the positive x-direction and the z-axis being the vertical axis is given by

$$\Delta c(x,y,z) = Q \, / \, [2\pi \, u \, \sigma_y(x) \, \sigma_z(x)] \times \exp\{-y^2/[2 \, \sigma_y^2(x)]\} \times$$

$$\exp \{-(z-h)^2/[2 \, \sigma_z^2(x)]\} \tag{2.20}$$

(Zenger 1998:86). The profile of the plume in the y-z plane perpendicular to the wind direction is a two-dimensional Gaussian distribution with standard deviations $\sigma_y(x)$ and $\sigma_z(x)$ increasing with the distance x away from the source. While these two parameters are theoretically proportional to the squareroot of the traveling time x/u (Seinfeld and Pandis 1998:944, eq. 18.70), they are in practice parameterized on the basis of comparisons of calculated and measured concentrations in order to obtain more realistic results. The Gaussian model according to the VDI Guideline 3782 (VDI 1992), which will be used in chapter 3, uses parameterizations of the form

$$\sigma_y(x) = F \, x^f, \quad \sigma_z(x) = G \, x^g \tag{2.21}$$

with parameters F, f, G, g given as a function of emission height and atmospheric stability class (VDI 1992:16; Zenger 1998:91). Modifications of the Gaussian profile in the vertical (z-) direction account for reflection of the plume at the ground (c = 0 for z < 0) as well as at the top of the mixing layer (the height of which is given as a function of the atmospheric stability class) and dry deposition at the ground. Wet deposition and a possible constant decay rate of the pollutant, which diminish pollutant concentrations in addition to dilution and dry deposition, are considered by an exponential decay factor. Furthermore, this model iteratively calculates the transport wind speed u as a function of the distance from the source as the mean of the vertical wind speed profile weighted with the vertical concentration profile of the pollutant. This makes the model particularly suitable to deal with traffic emissions, since the wind speed significantly increases with altitude close to the ground.

The basic formula (2.20) and the abovementioned modifications thereof apply to fixed values of wind speed, wind direction and atmospheric stability class. In order to calculate long-term (typically annual) means of the concentration around a source, combined frequency distributions of these parameters are used. For each parameter combination, the modified version of equation (2.20) is applied, and the resulting concentration fields are weighted with the respective frequency and summed up to yield the long-term (annual) mean concentration field.

One of the main factors influencing the accuracy of the concentrations predicted by Gaussian models is the parameterization of the standard deviations $\sigma_y(x)$ and $\sigma_z(x)$. The parameterizations given by different authors on the basis of various series of measurements of actual concentrations lead to differences of the annual

---

[13] The effective emission height is the sum of the stack height and the final value of the plume rise. Its determination is described in (VDI 1985). In the following, the term emission height will always refer to the effective emission height.

mean concentrations in the order of 30 %. Taking into account other parameter uncertainties and systematic simplifications of the Gaussian models, several authors state that the deviation of the calculated annual mean concentrations from the measured ones can be up to a factor 2 in both directions (Zenger 1998:107-109; VDI 1992:18).

These uncertainty estimates refer to concentrations at individual points around the source. In the present work, however, only the integral of the pollutant concentration (weighted with the population density) over a large area is considered (equation 2.16). Due to the conservation of the mass of the pollutant (including deposited mass and decay products), uncertainties of the pollutant concentrations at individual points can be expected to cancel each other out to a large degree. This holds at least to the extent that they are due to the modeling of the horizontal advection and diffusion of the pollutants, i.e. the parameterization of $\sigma_y(x)$ and the assumption of a spatially constant wind direction. The uncertainties due to the modeling of the vertical dispersion, on the other hand, remain unaffected by the horizontal area integration. Overall, the error in the fate factors and effective population densities due to the uncertainties of the Gaussian model alone can therefore be expected to be less than a factor of two.

This can be expected to hold even though the Gaussian dispersion model is applied here not only within 10 kilometers around the source (to which range the above error estimates presumably refer), but within a range of 100 kilometers (see chapter 3). Application within this larger range is still considered to lead to reasonable approximations of point values of the concentrations (VDI 1992:18-19), and hence the more so of their area-integrals. Overall, the uncertainty associated with determining the short-range contributions to the population exposures and fate factors of primary pollutants by using a Gaussian dispersion model are estimated to be a factor 2 in both directions with 95 % confidence, i.e. $\sigma_g^2 = 2$.

While Gaussian dispersion models are modified versions of a very specific solution of the differential equations describing the motion of air masses in the atmosphere and the distribution of air pollutants therein, a large variety of models exist that work with much less restrictive assumptions. In exchange for that, their equations can in general only be solved numerically. Many of these models first calculate wind fields for specified orographic and meteorological conditions.

With regards to the subsequent calculation of the dispersion of pollutants within those windfields, two basic approaches can be distinguished. Eulerian models solve the atmospheric advection-diffusion equation (Zenger 1998:35, eq. 3.28) within a coordinate system fixed relative to the earth. Lagrangian models, on the other hand, consider small air parcels (particles) that are moving along trajectories of the calculated mean wind field. The turbulent part of their motion is modeled as a Markov process, i.e. the fluctuations of the wind velocity around the mean velocity of the trajectory is modeled as the sum of a random component and a component correlated to previous fluctuations to consider the 'memory' effect due to the inert mass of the air. Lagrangian models consider a large number of air parcels (up to some 10,000) and determine the pollutant concentration fields from the distribution of these parcels (Zenger 1998:45-47, 125). Source-orientated models thereby consider trajectories starting at one common source point, while receptor-orientated models consider trajectories arriving at a common receptor point.

Besides the much more detailed description of the motion of air masses and pollutants compared to Gaussian models, both Eulerian and Lagrangian models offer the further advantage of being able to consider in detail the chemical transformation of the emitted pollutants and in particular the formation of secondary pollutants such as sulfate and nitrate aerosols or ozone. A Lagrangian dispersion model was used in the present work for this purpose as well as for the consideration of the transport of primary pollutants in the range beyond 100 kilometers away from the emission source. The main features of the model are described in section 3.4.2.1.

The purpose of using any of the above air pollution models is to determine the incremental pollutant concentration $\Delta c(\underline{x})$ around the considered emission source. Fate factors can then be determined according to equation (2.13). In order to calculate the effect factor and the effective population density according to equations (2.14) and (2.15), further information about the distribution of background concentrations $b(\underline{x})$ determining the shape of $d'(\underline{x}) = d'(b(\underline{x}))$ and the population density $\rho(\underline{x})$ are required. Methods for the Life Cycle Impact Assessment of health effects of air pollutants can be distinguished according to the level of detail at which such information is used. This determines whether the methods allow for a spatial differentiation of the impacts.

Existing methods to consider the spatial differentiation of fate factors and effective population densities: It was shown above that, for either linear exposure response functions or for marginal emission sources, the time characteristics of the emission do not play a role in terms of its incremental impact. The spatial characteristics of the emission source, i.e. its location and effective emission height, may have a significant influence, however. In order to discuss this influence, it should first be observed that the area-integral in equation (2.12) extends, in principle, over the whole surface of the Earth. For pollutants with atmospheric residence times in the order of one year or more, which is the typical timescale for interhemispherical mixing (Bliefert 1995:106), this maximum integration area will indeed have to be used. In that case, the integrand in (2.12) can be approximated by its worldwide average, and the impacts are therefore independent of the spatial source characteristics. A similar reasoning applies to pollutants with a residence time in the order of half a year which is sufficiently long for hemispherical mixing to be achieved.

A noticeable spatial differentiation of impacts is therefore only to be expected for pollutants with shorter atmospheric residence times in the order of hours or days rather than months or years. In this case, due to the decrease of $\Delta c(\underline{x})$ with increasing distance from the emission source, contributions to the area integral in (2.12) will mainly come from some vicinity around the source with an extension between some ten and some thousand kilometers (see sections 3.2 and 3.6.4 for more details). The incremental impact ΔD of the emission can then be expected to depend on the spatial characteristics of the emission, i.e. the emission height above ground and the geographical location of the emission source.

The effective emission height thereby directly influences the concentration $\Delta c(\underline{x})$ of primary pollutants at the ground within some 10 - 20 kilometers around the emission source. Emissions higher above the ground lead to lower concentrations and hence impacts. The position $\underline{x}_0$ of the emission source affects all three factors of the integrand in (2.12) either directly or via intermediary variables: The pollutant concentration $\Delta c(\underline{x})$ depends on the meteorological conditions around the source, such as wind speed, wind direction, atmospheric stability and possibly background concentrations of pollutants involved in chemical transformations of the considered pollutant. With regards to the population density $\rho(\underline{x})$ around the source, the spectrum of possibilities within one country is spanned by big cities on the one hand and rural areas on the other hand. On a larger spatial scale, countries of different average population densities, including unpopulated deserts or the sea, also need to be differentiated. In a similar way, areas of higher and lower background concentrations of the considered pollutant may be distinguished in terms of the slope $d'(b(\underline{x}))$ of a possibly nonlinear exposure response function.

Methods for the Life Cycle Impact Assessment of health effects of air pollutants can be distinguished according to whether they take the spatial differentiation of the impacts into account at all, and if so, at which level of detail. While a higher level of detail is in principle desirable, it needs to be balanced in terms of the required effort and time for data collection and calculations with the already demanding task to put together the data for the typically very large number of processes contained in a Life Cycle Inventory. This large number of processes also implies that each additional parameter to be collected is potentially associated with a high additional effort. The methods discussed in the following are located at different positions along this trade-off.

Early methods of Life Cycle Impact Assessment, such as the method of Critical Volumes (Habersatter 1990:107-111) or the CML method (Heijungs et al. 1992) assessed human health impacts solely on the basis of effect factors E (e.g. legal limit values) and the emitted mass M, whereas the large variability of the fate factors F over about four orders of magnitude due to the vastly different atmospheric residence times of the pollutants was disregarded (Jolliet et al. 1996:57), let alone the spatial variation of fate factors and effective population densities.

The abovementioned empirical method of Jolliet et al. (1997) provides spatially undifferentiated fate factors for air pollutants and does not allow for the calculation of spatially differentiated effective population densities. Hofstetter (1998) provides multimedia fate factors that, insofar as they concern air pollutants, are not differentiated with regards to emission height and in fact approximately represent the case of emissions from high stacks due to the assumption of an immediate vertical mixing of the pollutants. For emission sources located in the most densely populated part of Europe (population density 300 persons/km²), Hofstetter provides the effective population density as a function of the atmospheric residence time of the pollutants. The possibility of doing the same for emissions located in other parts of Europe is mentioned. In this way, a differentiation of the effective population densities for different countries in Europe would be possible. However, differentiations with regards to emission height and locations at a smaller scale (agglomerations versus rural areas within one country) are not provided.

Spadaro and Rabl (1999) provide damage factors for a number of air pollutants which refer to typical European conditions and are based on site-specific assessments for a large number of sites across Europe within the ExternE project (European Commission 1998). The spatial differentiation of the effects in terms of the location of the emission source either in or near a big city or at a rural site and in terms of the emission height is demonstrated in an exemplary way for a number of sites in France, but no general method for the characterization of sites in terms of the surrounding population density is provided.

A full consideration of the spatial differentiation of the impacts of air pollutants with regards to both emission height and location of the emission is possible by applying the method of Impact Pathway Analysis, as it was developed in the context of the ExternE project on external costs of electricity generation and transport (European Commission 1995, 1998; Friedrich et al. 1998), to Life Cycle Impact Assessment (Krewitt et al. 1998). Such a site-specific impact assessment requires the use of locally specific data on source location as well as population density distribution and meteorological conditions around the source. As is argued in more detail in section 3.1, it might not always be feasible to collect these data for all processes within an inventory, in particular not for the large number of sites to be considered for moving sources (vehicles) which are of particular interest here.

Therefore, a more generic method requiring fewer input data, but nevertheless allowing for a consideration of spatial differentiation of effective population densities and fate factors for the case of air pollutants with linear exposure response functions was developed and is presented in chapter 3. The level of spatial detail of this method for air pollutants is higher than that of the method of Hofstetter (1998) (which, on the other hand, also includes multimedia modeling for carcinogenic substances), but lower than that of the Impact Pathway Analysis according to Krewitt et al. (1998). The opposite applies with regards to the ease of application of these methods.

2.5.2.6
Photo-Oxidant Formation

This impact category comprises impacts on human health and vegetation, including agricultural crop production, due to the formation of photo-oxidants as secondary pollutants in the troposphere. The most important photo-oxidant is ozone (O_3), which is formed from nitrogen oxides (NO_x) and volatile organic compounds (VOC), mostly under conditions of high solar irradiation and atmospheric conditions characterized by only a small exchange of air masses both vertically and horizontally. Since high temperatures, in addition, favor the tropospheric ozone formation, it is also denoted as „summer-smog".

The photochemical reactions leading to the tropospheric ozone formation are complex and exhibit non-linear input-output relations. In particular, the reduction of either one of the precursor substances (NO_x or VOC) can, under certain conditions, lead to an increase in the ozone concentrations. A description of the chemical mechanisms of tropospheric ozone formation is beyond the scope of this chapter. A review is provided by Seinfeld and Pandis (1998:254-313).

In the context of LCA and other environmental assessments, either one of two types of proxy indicators for the relative contributions of the large number of different VOC to the formation of tropospheric ozone are frequently used. While the Maximum Incremental Reactivities (MIR, Carter 1994) refer to a worst case situation of maximum ozone creation, the Photochemical Ozone Creation Potentials (POCP, Derwent et al. 1998) are more representative of an average European situation of ozone formation. They will be used here as category indicators, as suggested by Udo de Haes et al. (1999:172). In both cases, contributions of the various primary pollutants to ozone formation are calculated relative to the contribution of the reference substance ethylene. POCP values therefore do not provide an indication of the absolute amount of ozone created by the emission of a given mass of a particular primary pollutant, e.g. by way of a fate factor F. In particular, they do not contain any information about the frequency of ozone episodes, which varies from year to year. The POCP values calculated by Derwent et al. (1998) refer to a modeled air parcel traveling for 5 days along a realistic trajectory leading from Vienna across Europe and France to the United Kingdom.

The relative POCP values equally apply to all effects of ozone, i.e. to damages to plants, agricultural crops and human health. As pointed out by Udo de Haes et al. (1999:172), the impacts on human health can be characterized further in a way comparable to the human toxicity indicators (see section 2.5.2.5). This further characterization step shall be pursued here since it will be of particular interest for the case study on transportation fuels to express all human health effects, be they due to carcinogens, particulate matter, ozone or other substances, in the same units (e.g. Years of Life Lost and Years Lived Disabled). In this way, their relative significance can be assessed by modeling the respective cause and effect chains up to the same endpoints rather than by pursuing a weighting on the basis of normative presuppositions.

For this purpose, fate factors for the ozone formation from NO_x and NMVOC, respectively, were estimated by Hofstetter (1998:307-311) as 1,5 E-6 m^2a/m^3 for both NO_x and NMVOC through an evaluation of the results of various versions of the EMEP (European Monitoring and Evaluation Program) model of ozone formation in Europe (Simpson 1992, 1993; Barrett and Berge 1996; Simpson et al. 1997). These fate factors are based on the assumption that ozone formation only occurs in the summer months (April-September). They represent European averages and do not allow for a spatial differentiation with regards to the location of the emission. Fate factors F_v for the individual VOC (index v) can be determined from the aggregate fate factor F_{NMVOC} for all NMVOCs by scaling the latter with the ratio $POCP_v/POCP_{NMVOC}$.

In combination with the effect factors for ozone listed in table 2.7 and an average population density for Europe of 80 persons/km^2 (including some part of the adjacent seas), the damage ΔD to human health in terms of Years of Life Lost, Years Lived Disabled and Disability Adjusted Life Years per mass M of emitted primary pollutants can be determined according to equation (2.4). The damage factors $\Delta D/M$ for the substances relevant to the case study in chapters 4 and 5 are listed in table 2.8 together with the corresponding fate factors and POCP values (Hofstetter 1998:307-313, 441-442). For a lack of emission data for individual substances within the group of NMVOC (see section 4.1.3 and 5.1.1.2), only aggregate NMVOC emissions will thereby be considered. For the vehicle emissions, these will be characterized by a country average POCP of 59,2 % (for the United Kingdom, Derwent et al. 1996, quoted after Hofstetter 1998:312), which is a plausible choice, since road transport is the main contributor to NMVOC emissions. Sector-specific POCPs are furthermore available for the supply of natural gas (POCP = 1,1 %) and liquid fossil fuels (POCP = 2,2 %; see table 2.8).

Table 2.8. Damage factors for human health effects through ozone formation for a selection of primary pollutants and pollutant groups relevant for the case study in chapters 4 and 5 (Hofstetter 1998:307-313, 441-442)

primary pollutant p	POCP [%]	Fate Factor (p → O₃) [m²a/m³]	YLD / M [a/kg]	YLL / M [a/kg]	DALY / M [a/kg]
NO$_x$ (as NO$_2$)		1,5 E-6	0,9 E-6	0,4 E-6	1,3 E-6
NMVOC (sum)	59,2[a]	1,5 E-6	0,9 E-6	0,4 E-6	1,3 E-6
methane	0,6	1,5 E-8	0,9 E-8	0,4 E-8	1,3 E-8
natural gas leaks	1,1[b]	2,8 E-8	1,6 E-8	0,8 E-8	2,4 E-8
oil transport and refining	2,2[c]	5,6 E-8	3,2 E-8	1,6 E-8	4,8 E-8

POCP Photochemical Ozone Creation Potential, *YLD* Years Lived Disabled, *YLL* Years of Life Lost, *DALY* Disability Adjusted Life Years, *M* mass of emitted pollutant

[a] based on a detailed emission inventory for the UK (Derwent et al. 1996, quoted after Hofstetter 1998:312).

[b] calculated by combining the composition of natural gas in Germany according to Fritsche et al. (1998) with the POCP values of Derwent et al. (1998).

[c] A ratio of 2 between the POCP values for oil transport and refining and for natural gas leaks was used (Hauschild and Wenzel 1997:104).

2.6
Interpretation

While the main outcome of the Life Cycle Inventory (LCI) phase and the Life Cycle Impact Assessment (LCIA) phase of a Life Cycle Assessment (LCA) study are the respective quantitative results, the purpose of the interpretation phase is an appraisal of the quality of those results and the drawing of conclusions. The interpretation phase comprises three elements:

1. The identification of significant issues of the product system. Significant issues are essential contributions to the LCI and LCIA results for the individual life cycle stages.
2. An evaluation of the LCI and LCIA results in terms of completeness, sensitivity to particular input parameters in order to check the reliability of the results, and the consistency of the assumptions, methods and data with the goal and scope of the study.
3. Conclusions and recommendations, which include a reporting of the significant issues (ISO 1999b).

3 Site-Dependent Impact Indicators for Human Health Effects of Airborne Pollutants[1]

3.1 Introduction

The marginal health impact of a given emission of an airborne pollutant depends upon

- the effective emission height[2], which influences the pollutant concentrations at the ground
- the population density around the emission source (which differs between large cities and rural districts, and on a larger spatial scale between more or less densely populated countries)
- the meteorological conditions around the source, which affect the dispersion of the pollutants
- the background concentration of the pollutant in the case of a nonlinear exposure-response function (see section 2.5.2.5).

The question to be addressed here is how this spatial differentiation of the marginal impacts of emissions of airborne pollutants can be considered within Life Cycle Impact Assessment (LCIA). One possibility is to carry out a *site-specific* impact assessment, which requires the location of the emission source to be known exactly and uses locally specific data on the population density distribution and the meteorological conditions (i.e. wind speed, wind direction, atmospheric stability) around the source. In particular, the application of the site-specific Impact Pathway Methodology within LCIA has been suggested by Krewitt et al. (1998). In order to simplify this task, the software tool EcoSense has been developed (IER 1998). This tool contains databases on population densities and on the background concentrations of some air pollutants. However, at present, it still requires site-specific meteorological data (wind speed, wind direction, atmospheric stability) to be added by the user. The collection of these data is either time consuming, or the cost for purchasing them on a commercial basis is rather high[3]. Given the large number of emission sources that typically need to be con-

---

[1] Parts of this chapter were first published in (Nigge 1998).

[2] The effective emission height is the sum of the stack height and the final value of the plume rise. Its determination is described in (VDI 1985). In the following, the term emission height will always refer to the effective emission height.

[3] The German meteorological service (Deutscher Wetterdienst), for example, sells one set of meteorological data at a cost of about 100 Euro.

sidered in a Life Cycle Inventory, a site-specific impact assessment for all of them might not always be feasible. There may also be cases where the location of the emission source is not known exactly, but only in terms of some general characteristics such as urban, suburban or rural. Such a more generic characterization of emission sites may even be desirable, for example in the case study presented here (see section 4.1.1).

For these reasons, it has been proposed to treat possible spatial variations of the impacts of pollutants in a more generic way by considering a limited number of types of emission situations (Potting and Hauschild 1997). This approach is called *site-dependent* Life Cycle Impact Assessment (as opposed to site-specific). The types of emission situations are called *generic spatial classes*. Their use within LCA has been recommended by the Second Working Group on Life Cycle Impact Assessment of SETAC-Europe (Udo de Haes et al. 1999:73). For each generic spatial class, an impact assessment would be carried out using input data (e.g. receptor distributions, meteorological data) that can be considered as characteristic for the class.

The site-dependent approach to LCIA can be seen as complementary to a site-specific approach. Due to the lower demand for additional input data (i.e. input data that are not usually contained in Life Cycle Inventories today), the generic spatial classes can be attributed to all relevant emissions in an inventory. Should a further spatial differentiation be required to arrive at an overall conclusion, a site-specific impact assessment can be carried out for those emissions contributing the most to the overall impact. In this sense, the generic spatial classes can be seen as a screening tool to determine whether and for which emissions a further site-specific impact assessment may eventually be required to arrive at a definite overall conclusion.

As will be shown, the degree of spatial differentiation is highest for primary pollutants from traffic emissions due to their low emission heights. The approach of defining generic spatial classes is particularly suitable to deal with traffic emissions. On the one hand, a site-specific assessment would require even more locally specific input data in the case of a mobile source. On the other hand, the overall impact of a moving source is an average over the site-specific impacts along its path, and therefore less sensitive to the site-specific conditions. Furthermore, even if the location of the source is not known exactly, some approximate information about the site of the traffic emission may be gained from the inventory data: Emission factors for vehicles are typically differentiated between driving in cities, on rural roads or on highways. This approximate information may then be expressed in terms of the generic spatial classes.

The method presented in this chapter, while being applicable to arbitrary emission heights, was developed with a particular view on traffic emissions and the application to LCAs dealing with transportation fuels (see case study in chapters 4 and 5). Therefore, the method follows the idea of capturing the spatial differentiation of the health impacts of airborne pollutants through generic spatial classes. With regards to the operationalization of the idea of generic spatial classes, three questions arise:

1. How can suitable generic spatial classes be defined?
2. How can average impacts for the classes be calculated?
3. How can the variability of the impacts *within* the classes be estimated?

The relevance of the third question is that estimates of the intra-class variability of the impacts can be used for sensitivity analyses: if the intra-class variability leads to changes in the overall conclusions, a further site-specific assessment is required. Therefore, the variability estimates facilitate the use of the generic spatial classes as a reliable screening tool.

The method presented here addresses these questions for the case of *primary* airborne pollutants for which a *linear exposure-response function* can be assumed. For emissions of nitrogen oxides and sulfur dioxide, the formation of secondary nitrate and sulfate aerosols is also considered. The question of the spatial differentiation of the impacts of ozone formed from nitrogen oxides and volatile organic compounds is a topic for further research.

The assumption of a linear exposure-response function (at the population level) is made so that emission sites, to begin with, only need to be classified in terms of their surrounding population density, but not also in terms of the background concentrations in their vicinity. The possibility of a classification with regards to both parameters is a topic for further research. The linearity assumption represents a limitation of the method. Nevertheless, it covers the most important pollutants for the case study on natural gas and Diesel vehicles in chapters 4 and 5. In fact, *all* exposure-response functions used in a recent major study on the external costs of transport were of the linear type (Friedrich et al. 1998). The assumption of a linear exposure-response function is commonly made for carcinogenic substances. Furthermore, respiratory health effects of air pollutants such as primary particulate matter, carbon monoxide, sulfur dioxide, nitrogen dioxide as well as secondary sulfate and nitrate aerosols and ozone can also, in good approximation, be described by linear exposure response functions (see section 2.5.2.5, tables 2.6 and 2.7).

In order to address the three research questions formulated above, the method presented here uses a verified statistical reasoning as its main underlying idea (sections 3.2 and 3.5). Under the assumption of a linear exposure-response function, the background concentrations of pollutants do not need to be considered. Furthermore, the emission height is kept as a continuously variable parameter, since its determination is relatively easy. This leaves two factors to be considered for the definition of the generic spatial classes, namely the population density (section 3.3) and the meteorological conditions (section 3.4). Exemplary results showing the spatial differentiation of the health impacts of airborne pollutants and its representation by means of generic spatial classes are presented and discussed in section 3.6.

3.2
Statistical Calculation of Class Averages

The method presented here uses statistical arguments to calculate average impacts for the generic spatial classes (which will be defined in section 3.3). It may be recalled from section 2.5.2.5 that, in the case of a linear exposure-response function, the incremental damage ΔD_i to human health of a given emission of a particular pollutant at a specific site i is independent of the background concentration of the pollutant and of the time characteristics of the emission. It can be calculated according to equation (2.16) as

$$\Delta D_i = E \, M \, I_i \tag{3.1}$$

with

E   effect factor, i.e. the constant slope of the exposure-response function
M   total emitted mass
I_i incremental population exposure per mass of emitted pollutant, i.e. $I_i = \Delta PE_i/M$ (equation 2.18). In two-dimensional polar coordinates (r,φ) around the emission source within a suitable cartographic projection of the earth's surface, this can be written as

$$I_i = 1/Q \int_0^R r \int_0^{2\pi} \Delta c_i(r,\varphi) \, \rho_i(r,\varphi). \, d\varphi \, dr. \tag{3.2}$$

In principle, the area-integration in (3.2) is over the whole surface of the earth. In practice, the upper integration limit R will be chosen such that the major part (e.g. 95%) of the contributions to I_i is covered. For the case of power plant emissions of pollutants with a lifetime in the order of one day (particulate matter PM 10, SO_2 and NO_x), it has been shown that this requires R to be in the order of several thousand kilometers (European Commission 1995). In general, the required upper integration limit R will depend on the atmospheric residence time of the pollutants as well as on the emission height, with higher values of both parameters leading to higher required values of R. This will be discussed in more detail in section 3.6.4.

The idea of the statistical approach presented here is to identify classes C of emission situations i such that the values of I_i are similar for the sites within each class, yet different among the classes in a statistically significant way. Furthermore, the classes should be few in number in order to ensure that the classification can be used easily.

For the further discussions, it will be suitable to split the radial integral in (3.2) into a short-range contribution $I_{i,near}$ ranging from 0 to 100 kilometers and a long-range contribution $I_{i,far}$ ranging from 100 kilometers to R. It will be shown that the short-range contribution to the population exposure can significantly vary when the emission source is moved on a scale of some ten kilometers. For example, the value of $I_{i,near}$ strongly depends on whether the emission source is located in a city or in a less densely populated area some ten kilometers away from the city. The long-range contribution $I_{i,far}$, on the other hand, only varies on a larger spatial scale. As will be shown, it can be assumed to be constant for all emission sites within one country in a first approximation (sections 3.4.2.2 and 3.6.3). Therefore, only the short-range part $I_{i,near}$ of the integral I_i will be discussed in the following[4].

The main reason for the difficulties of evaluating $I_{i,near}$ in a site-specific way are associated with the angular variable φ in (3.2), because the angular dependence of

---

[4] A quantitative justification for choosing r = 100 kilometers as the border between short-range and long-range contribution is given in section 3.3.2.

the concentration is determined by the distribution of wind directions, which can vary significantly on a scale of some ten kilometers (Zenger 1998:51-52). Therefore, a desirable simplification would be to eliminate the angular variable φ. This can be achieved by looking at the average impact for a class C of emission sites and applying some statistical considerations. If the class C consists of N_C emission sites i, at each of which an identical emission source is located, then the average impact of these emission sources is obtained by using the average value

$$<I_i>_C = 1/N_C \sum_{i \in C} I_i \tag{3.3}$$

of the integral (3.2) instead of a specific value I_i in (3.1). After substituting (3.2) into (3.3) and reversing the order of summation and integration, the integrand $<\Delta c_i(r,\varphi) \times \rho_i(r,\varphi)>_C$ can be simplified as follows. Assuming a given effective release height, the variation of the concentration profiles $\Delta c_i(r,\varphi)$ for the different locations i is due to the different meteorological conditions (distributions of wind direction, wind speed and atmospheric stability). It seems plausible to assume that

(A1) the variation of meteorological conditions (in particular the wind direction distributions) and the variation of population density with the emission site i are statistically uncorrelated, i.e. $\rho_i(r,\varphi)$ and $\Delta c_i(r,\varphi)$ are uncorrelated with regards to variation of i for each fixed (r,φ).

In less abstract terms this means that, at some sites, the predominant wind direction is towards areas with higher population densities, at some sites towards areas with lower population densities, with no overall correlation between wind direction and direction of higher population density. In that case,

$$<\Delta c_i(r,\varphi) \times \rho_i(r,\varphi)>_C = <\Delta c_i(r,\varphi)>_C \times <\rho_i(r,\varphi)>_C . \tag{3.4}$$

If the sites within class C are geographically spread over a large area, it can be further assumed that

(A2) the second factor in equation (3.4) is independent of the angle φ, i.e., on average, there is no preferred direction for the spatial variation of ρ.

In that case, it can also be written as

$$<\rho_i(r,\varphi)>_C = <\rho_i(r)>_C \tag{3.5}$$

with

$$\rho_i(r) \equiv 1/2\pi \int_0^{2\pi} \rho_i(r,\varphi) \, d\varphi \tag{3.6}$$

being the *radial* population density distribution around site i. For the generic spatial classes to be defined in section 3.3, it turns out that the statistical assumptions (A1) and (A2) are fulfilled in very good approximation. This will be shown in section 3.5. Using (3.4) and (3.5) and introducing

$$\Delta c_i(r) \equiv 1/2\pi \int_0^{2\pi} \Delta c_i(r,\varphi) \, d\varphi \tag{3.7}$$

leads to

$$I_C \equiv <I_i>_C = \int_0^{100\ km} <\rho_i(r)>_C <\Delta c_i(r)>_C\ 2\pi r\ dr\ +\ I_{far}\ =\ I_{C,near} + I_{far}. \quad (3.8)$$

By using equation (3.8), the task of calculating average impacts for the generic spatial classes can be reduced to calculating class-averaged radial population density distributions $<\rho_i(r)>_C$ and class-averaged radial concentration profiles $<\Delta c_i(r)>_C$. Furthermore, equation (3.8) suggests using the radial population density distributions $\rho_i(r)$ and the radial concentration profiles $\Delta c_i(r)$ within a distance of up to 100 kilometers away from the sites i as criteria for the definition of the generic spatial classes. The operationalization of these criteria and the calculation of $<\rho_i(r)>_C$ and $<\Delta c_i(r)>_C$ is described in sections 3.3 and 3.4.

3.3
Classification of Emission Sites in Terms of Population Density

3.3.1
Classification Scheme

The classification of emission sites in terms of their radial population density distribution in the range between 0 and 100 kilometers is demonstrated here for the case of Germany. In order to operationalize the classification, an existing official classification of settlement structures (BBR 1998a) was used[5]. Areas at three different hierarchical levels, which are in part administrative units and in part analytical units, are classified essentially according to their population density. The top level of the hierarchy consists of 100 regions (Raumordnungsregionen), which are analytical units into which the German states (Länder) are subdivided. These are classified into agglomerated regions (n = 32), urbanized regions (n = 42) and rural regions (n = 26) according to the criteria listed in table 3.1.

Each region contains a number of districts (Kreise). In the terminology for ad-

Table 3.1. Classification of regions in Germany according to their settlement structure (BBR 1998a)

class of regions	criteria for classification
(I) agglomerated regions	$\rho > 300\ km^{-2}$ OR contain city with > 300,000 inhabitants
(II) urbanized regions	$\rho > 150\ km^{-2}$ OR ($\rho > 100\ km^{-2}$ AND contain city with > 100,000 inhabitants)
(III) rural regions	otherwise

ρ population density

---

[5] Based on a similar classification for Western Europe (Schmidt-Seiwert 1997), the method presented here can be extended to countries other than Germany.

Table 3.2. Classification of districts (NUTS level 3 units, Kreise) according to their settlement structure (BBR 1998a)

	NUTS level 3 class	criteria for classification
(1)	central city	city with > 100,000 inhabitants
(2)	highly densified	$\rho > 300$ km^{-2}
(3)	densified	$\rho > 150$ km^{-2}
(4)	rural with higher density	$\rho > 100$ km^{-2}
(5)	rural with lower density	$\rho < 100$ km^{-2}

ρ population density

ministrative units (Nomenclature des Unités Territoriales Statistiques, NUTS) of the Statistical Office of the European Communities (Eurostat 1995), they represent level 3. At the medium level of the hierarchical classification, the districts are grouped into five classes basically according to their population density. The detailed classification criteria are shown in table 3.2.

The combinations of region classes (I-III) and district (or NUTS level 3) classes (1-5) will be called *combined* district (or NUTS level 3) classes in the following. Out of the 15 possible combinations, only 9 do actually occur in sufficient numbers. Using the notation "(region class number, district class number)", these are: (I,1), (I,2), (I,3), (I, 4 and 5 merged), (II,1), (II,3), (II, 4 and 5 merged), (III,4), (III,5). Their geographical distribution is shown in fig. 3.1.

A district (Kreis) in Germany may consist of a number of municipalities (Gemeinden). In the NUTS terminology, these smallest administrative units represent level 5, since there are no level 4 units in Germany. They form the bottom level of the threefold hierarchical classification used here. They are essentially grouped into two classes (a) and (b) according to whether they fulfil the central functions of a city or not.

In the case of the combined classes (I,1) and (II,1), this distinction does not apply since these municipalities are identical with the cities at the NUTS level 3. Instead, the municipalities belonging to class (I,1) are subdivided according to whether they have (a) more or (b) less than 500,000 inhabitants. Municipalities belonging to class (II,1) are not subdivided further. On the whole, there are 9×2 - 1 = 17 possible combinations of the classes at the region level (I-III), the NUTS level 3 (1-5) and the NUTS level 5 (a-b). In the following, these shall be referred to as *combined* municipality (or NUTS level 5) classes. A notation analogous to the one for the combined district classes (e.g. (I,1,a)) will be used. The overall hierarchical classification scheme is shown in table 3.3 together with the numbers of municipalities that belong to each class.

3.3.2
Radial Population Density Distributions

Classes of emission sites could potentially be identified at each of the three levels of this hierarchical classification scheme. In order to calculate $<I>_{C,near}$ for the

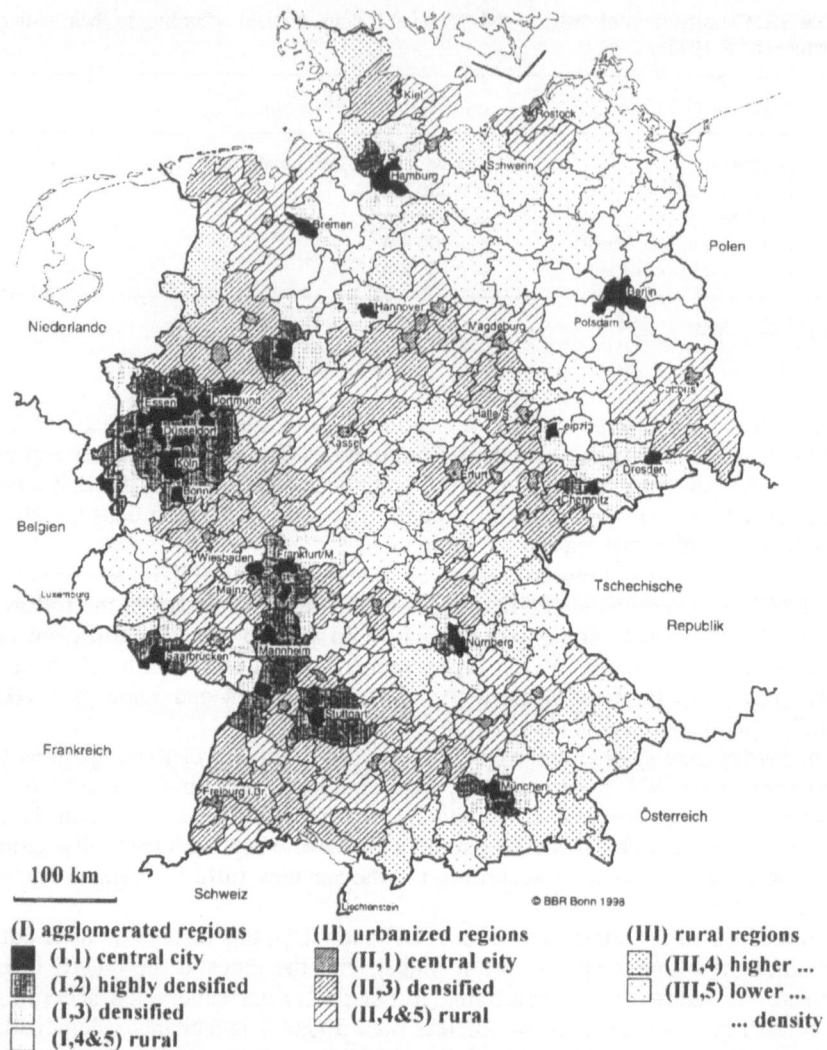

(I) agglomerated regions
- ■ (I,1) central city
- ▦ (I,2) highly densified
- □ (I,3) densified
- □ (I,4&5) rural

(II) urbanized regions
- ▨ (II,1) central city
- ▨ (II,3) densified
- ▨ (II,4&5) rural

(III) rural regions
- ▦ (III,4) higher ...
- □ (III,5) lower ...
- ... density

Fig. 3.1. Geographical distribution of the combined district classes in Germany (BBR 1998a)

classes at each level, the radial population density distribution $\rho_i(r)$ was determined for each municipality (Gemeinde) in Germany (i = 1..14615). The data regarding their areas, numbers of inhabitants, the spatial coordinates of their centers and their classification according to the scheme described above were obtained from (BBR 1998b).

The radial scale was divided into intervals of 10 kilometers. For each annulus of that width around site i, the total population of all municipalities in Germany with their center lying within the annulus was divided by their total area. The increment of 10 kilometers was chosen corresponding to the spatial resolution of

Table 3.3. Hierarchical classification of settlement structures in Germany at three levels. The number of municipalities (NUTS level 5 units) belonging to the respective classes is indicated in pointed brackets[a].

class of regions	combined district class		combined municipality class	
(I) agglomerated region {3420}	(I,1)	central city {53}	(I,1,a)	> 500,000 inhabitants {12}
			(I,1,b)	< 500,000 inhabitants {46}
	(I,2)	highly densified {782}	(I,2,a)	city {264}
			(I,2,b)	other {516}
	(I,3)	densified {1107}	(I,3,a)	city {185}
			(I,3,b)	other {919}
	(I,4&5)	rural {1478}	(I,4&5,a)	city {161}
			(I,4&5,b)	other {1317}
(II) urbanized region {6503}	(II,1)	central city {29}	(II,1,a)	city {31}
	(II,3)	densified {3376}	(II,3,a)	city {689}
			(II,3,b)	other {2685}
	(II,4&5)	rural {3098}	(II,4&5,a)	city {509}
			(II,4&5,b)	other {2589}
(III) rural region {4692}	(III,4)	rural with higher density {1847}	(III,4,a)	city {175}
			(III,4,b)	other {1672}
	(III,5)	rural with lower density {2845}	(III,5,a)	city {338}
			(III,5,b)	other {2507}

[a] Some of the numbers in the third column do not exactly add up to the numbers in the second column, since a few {7} cities with more than 100,000 inhabitants have been assigned to classes (2) and (3) at the NUTS level 3, deviating from the criteria given in table 2, but are nevertheless counted as if they belonged to class (1) in the third column.

the input data, given that the linear extension of a municipality (expressed as the radius of a circle with the same area) is typically in the range of a few up to 10 kilometers. In particular, the inner circle with a radius of 10 kilometers covers, in most cases, all of the area of the municipality unit with its center at $r = 0$. Otherwise, the population density of the municipality at the origin would be attributed to an area which is too small. In the case of big cities with linear extensions of more than 10 kilometers (class (I,1,a)), a site-specific adaptation of the radius of the inner circle could be carried out, which was not done here, however.

For the calculation of $<I>_{C,near}$, values of the radial population density distributions determined in this way are used for the range between 0 and 100 kilometers. Hence, for the municipalities within 100 kilometers of the borders of Germany, this procedure amounts to partially extending the radial population density distri-

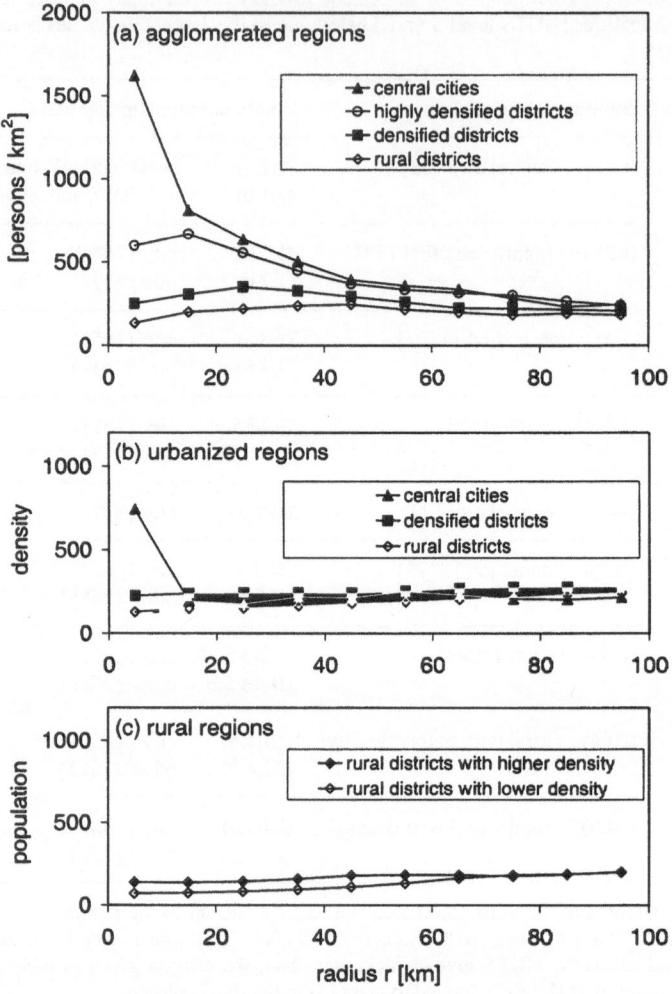

Fig. 3.2. Radial population density distributions in steps of 10 kilometers for the municipalities (NUTS level 5 units) in Germany, averaged within each of the 9 combined district classes belonging to (a) agglomerated regions, (b) urbanized regions and (c) rural regions. The data points are plotted in the middle of each interval of width 10 km. The lines connecting the data points are only meant to enhance graphical clarity but do not represent interpolations.

bution within Germany beyond its borders. This approximation is considered to be sufficient for the determination of class averages. Fig. 3.2 shows the radial population density distributions averaged within each of the 9 combined NUTS level 3 classes. The values of the radial population density for the combined NUTS level 3 classes only differ from the country wide average (about 230 persons per km^2, BBR 1998a) for distances of less than about 100 kilometers. This provides the quantitative justification for choosing a radius of $r = 100$ kilometers as the border between the short-range and the long-range contribution to the population exposure, and for approximating the long-range contribution by a constant value for all emission sites within one country.

The structures that can be discerned within the range of 0 and 100 kilometers are determined by both the region class and the district (NUTS level 3) class. This is the reason why the hierarchical classification of settlement structures at different spatial scales (region, district, municipality) is important. Figure 3.2(a) confirms that cities with more than 500,000 inhabitants within agglomerated regions (I,1) have by far the highest population density, while the decreasing population density of the classes (I,2), (I,3) and (I,4&5) goes along with an increasing distance from the areas of highest population density within the region. Within urbanized regions (class II, fig. 3.2(b)), only the central cities (II,1) show a narrow peak in the density distribution, while the other two classes show rather uniform distributions with values of the population density around and below the country wide average of about 230 persons per km^2, respectively. Rural regions (figure 3.2(c)) are characterized by relatively uniform distributions with values below the country-wide average, with a slight and very broad dip extending between 0 and about 50 kilometers in the case of class (III,5). A further subdivision of the 9 combined NUTS level 3 classes into the 17 combined NUTS level 5 classes does not, in general, lead to significant differentiations with regards to the values of $<I>_{C,near}$. This will be discussed in detail in section 3.6.1.

3.4
Atmospheric Dispersion of Pollutants

3.4.1
Dispersion of Primary Pollutants in the Short Range

3.4.1.1
Gaussian Dispersion Model

In order to calculate the short range contribution $<I_i>_{C,near}$ for a class C of emission sites according to equation (3.8), the second factor to be determined besides the class average radial population density distribution $<\rho_i(r)>_C$ is the class-averaged radial concentration profile $<c_i(r)>_C$ for values of the radius r between 0 and 100 kilometers. Concentrations of primary pollutants, which are either chemically inert (e.g. particulate matter) or for which only first-order chemical reactions with constant reaction rates need to be considered as decay processes (e.g. reaction of acetaldehyde with hydroxyl radicals) can be calculated with Gaussian dispersion models. Furthermore, since the spatial range of applicability of these models extends up to 100 kilometers (VDI 1992), they are suitable for the present purpose.

The Gaussian dispersion model used here fulfils the requirements of the Guideline 3782 of the German engineering society VDI (1992). Total reflection at the top of the mixing layer is assumed. The default values given in the Guideline for the height of the mixing layer as a function of the atmospheric stability class were used (stability classes I and II: 250 m, III_1 and III_2: 800 m, IV and V: 1100 m). The model calculates the transport wind speed as a function of the distance from the source by an integration of the vertical wind speed profile weighted with the vertical concentration profile. It allows for the incorporation of dry deposition, wet deposition and first-order chemical transformations. For the calculations, a computer program provided by Janicke (1998) was used.

A limitation of the dispersion model used here is that it assumes a flat terrain around the source, i.e. the effect of obstacles such as in street canyons or topographical features such as valleys or hills is not taken into account. With regards to topógraphical features, this simplifying assumptions seems unavoidable as long as one is dealing with classes of sites rather than specified individual sites. Street canyons could, in principle be considered for classes consisting of urban sites. However, their influence on the impacts of pollutants with linear exposure response functions turns out to be negligible, as will be shown in section 3.6.4.

For the purpose of dispersion modeling, emissions from vehicles can be represented as line sources. In order to calculate $<I_i>_C$ for traffic emissions, it is assumed that all points of the line source are located in an area that belongs to class C. Since all sites within class C are characterized by the *same* radial population density distribution and the *same* meteorological conditions within the present method, each infinitesimal segment on the line source contributes the same amount to $<I_i>_C$. For the purpose of evaluating $<I_i>_C$, the line source can therefore be condensed into a point source and treated in the same way as emissions from stationary sources.

The effective emission height of the point source representing traffic emissions was assumed to be 5 meters as an estimated average for passenger cars, commercial vehicles and trucks, taking into account that there might be some plume rise due to the mechanical impulse or the temperature of the fumes. Possible time correlations between typical variations of traffic volume and meteorological conditions (e.g. atmospheric stability) in the course of a day were not considered.

3.4.1.2
Generic Meteorological Data

Emission processes considered in Life Cycle Assessments usually cannot be specified very precisely in terms of the calendar time when they occur. The calendar time is sometimes specified in terms of the calendar year, but shorter time-periods are usually not considered[6]. Therefore, only annual mean pollutant concentrations are considered in the following. In order to calculate the annual mean concentration around an emission source with a constant rate of emission, the Gaussian model used here requires as meteorological input data the combined frequency distribution w_{isud} of the short-term (e.g. hourly) values of atmospheric stability class (index s), wind speed u and wind direction (index d) at the site i. All three parameters (s, u, d) are thereby discretized into a finite number of values. The Gaussian model used here works with six atmospheric stability classes (i.e. $s = 1...6$), 15 intervals of width 1 m/s of the short-term wind speeds u and 36 angular sectors for the wind direction ($d = 1...36$) represented in the following by the direction φ_d towards which the wind is blowing. The annual mean concentration $\Delta c_i(r,\varphi)$ is then calculated according to

---

[6] see Hofstetter (1996) for a detailed discussion.

$$\Delta c_i(r,\varphi) = \sum_s \sum_u \sum_d w_{isud} \, \Delta c_{su}(r,\varphi-\varphi_d) \tag{3.9}$$

It is important to note that, relative to the direction φ_d and for fixed s and u, the concentration field Δc_{su} is always the same, i.e. it only depends on $\varphi-\varphi_d$. This is a consequence of the assumption of a flat terrain, which 'looks the same' in all directions φ. In Cartesian coordinates with the positive x-axis in the direction $\varphi = \varphi_d$, Δc_{su} is given by equation (2.20)[7] with $z = 0$.

According to equations (3.7) and (3.8), only the angular average $\Delta c_i(r)$ of the pollutant concentration is of interest here. Inserting (3.9) into (3.7), carrying the angular integration through the summations over s, u and d and using a variable substitution $\varphi' = \varphi-\varphi_d$ in each integral leads to

$$\Delta c_i(r) = \sum_s \sum_u w'_{isu} \left[1/2\pi \int_0^{2\pi} \Delta c_{su}(r,\varphi') \, d\varphi' \right] \tag{3.10}$$

with

$$w'_{isu} = \sum_d w_{isud} \tag{3.11}$$

being the combined frequency distribution of atmospheric stability class and wind speed at the site i. Since w'_{isu} is *independent* of the frequency distribution of the wind *directions*, this also applies to $\Delta c_i(r)$. This eliminates much of the site-specificity of the meteorological data, which is brought about by the circumstance that wind directions close to the ground can vary on a scale of some ten kilometers due local topographical conditions (Zenger 1998:51-52).

In order to calculate $\Delta c_i(r)$, the angular integral in equation (3.10) was evaluated numerically by approximating it by a discrete sum over 36 equal intervals of φ'. The remaining task is to determine combined frequency distributions w'_{isu} of the atmospheric stability class and the wind speed. These distributions are less locally variable and can be correlated well with a simple meteorological parameter, the annual mean wind speed, in the following way. First, for a given value u_{mi} of the annual mean wind speed at site i, a frequency distribution of short-term wind speeds u can be derived using the circumstance that it can be approximated by a Weibull distribution

$$f_i(u) = k/a_i \, (u/a_i)^{k-1} \exp \left[- (u/a_i)^k \right] \tag{3.12}$$

where k is called ‚shape-parameter' and a_i ‚scale-parameter'. The scale-parameter is related to the annual mean wind speed u_{mi} through

$$u_{mi} = a_i \, \Gamma(1 + 1/k). \tag{3.13}$$

With regards to the shape-parameter k, an evaluation of wind measurements for 41 stations in Western Germany over a period of 10 years found its values to vary in the range between 1,2 and 2,3, with higher values (more narrow distributions) at the coast and in some elevated areas and lower values (broader distributions) elsewhere (Christoffer and Ulbricht-Eissing 1989). The average value of k over all stations (k=1,7) was used here, since its standard deviation is only 17 % and $\Delta c_i(r)$ is not very sensitive to k anyway. In a second step, a set of regression formulae derived from meteorological observations over a period of 5 years for 30 stations in Germany was used to determine the combined frequency distribution w'_{isu} of

---

[7] More precisely, a modification of equation (2.20) was used here (see section 2.5.2.5).

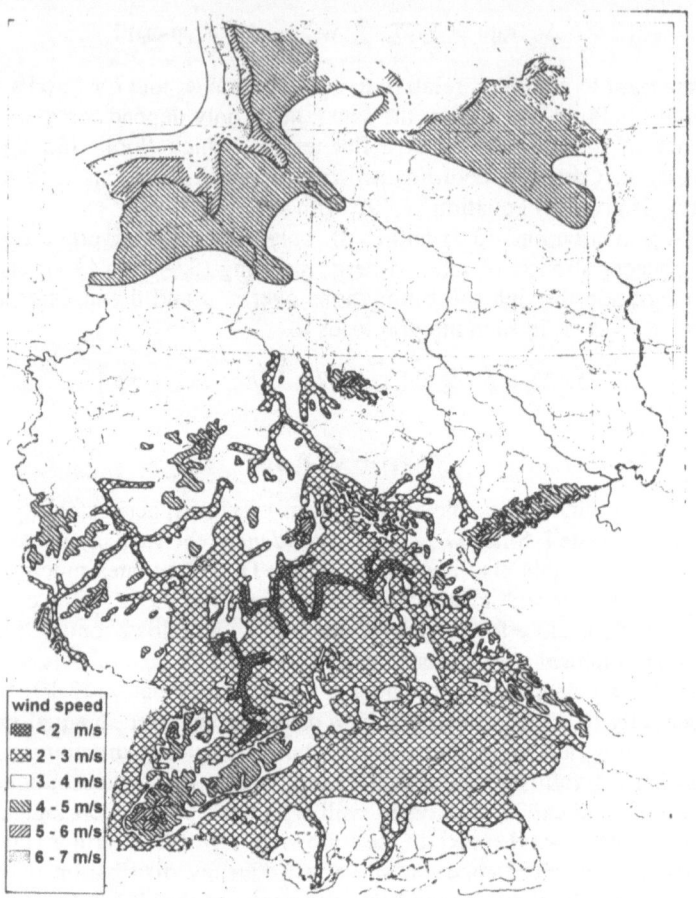

Fig. 3.3. Zones of different annual mean wind speeds at 10 meters above ground in Germany (Gerth and Christoffer 1994:69). © Deutscher Wetterdienst, Offenbach

wind speeds and stability classes from the frequency distribution $f_i(u)$ of the wind speeds. The error associated with this procedure is in the range of 10 % (Manier 1971).

Therefore, the annual mean wind speed in a circle of radius 100 kilometers around the emission site can be used as a second criterion for the definition of suitable generic spatial classes. For the purpose of a site-dependent impact assessment, it is convenient to discretize the annual mean wind speeds in intervals of 1 m/s. In the case of Germany, this essentially leads to three classes of annual mean wind speeds in the ranges of 2-3 m/s, 3-4 m/s and 4-5 m/s, respectively. As fig. 3.3 shows, they cover most of the area of Germany. Depending on the degree of knowledge about the site of an emission source, the appropriate interval of the annual mean wind speed can either be determined from fig. 3.3, or a default value of 3,5 m/s for Germany can be used.

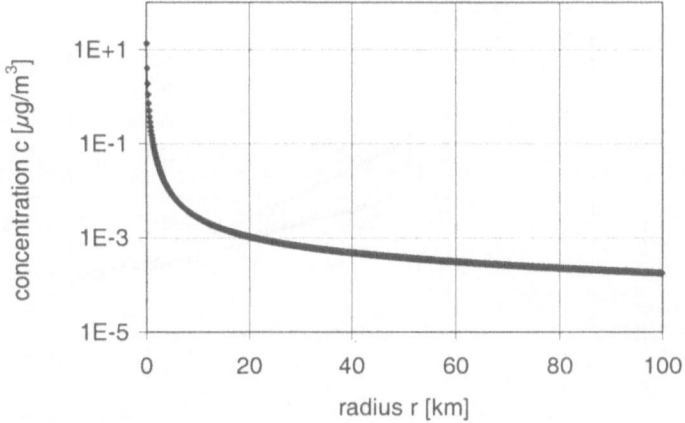

Fig. 3.4. Radial concentration profile for a point source of Diesel particles (dry deposition velocity $v_d = 0,065$ cm/s, wet deposition velocity $v_w = 0,12$ cm/s) with an effective emission height of 5 meters above ground for an emission rate of 1 kg/h and an annual mean wind speed of 3,5 m/s (default value for Germany)

Some limitations apply to this way of determining the annual mean wind speed. Fig. 3.3 is based on wind speed measurements at 41 stations in western Germany which can be considered to be representative for the regions or landscapes in which they are located in terms of topographical structure and surface roughness. The influence of particular local variations of these parameters, such as topographic features in the more hilly southern areas of Germany or the increased roughness length within cities, on the wind speed is not represented. While the effects of particular local topographic features on the wind speed are likely to cancel each other out over a transport distance of 100 kilometers, the reduction of the wind speed within cities due to the increased roughness length leads to systematically higher pollutant concentrations. This effect can, in principle, be considered for those spatial classes consisting of urban sites. However, since the annual mean wind speed turns out to have a smaller influence on the values of $<I_i>_{C,near}$ compared to the population density, the effect can be considered not to be of major relevance and was therefore not considered here.

3.4.1.3
Pollutant Concentrations

Pollutant concentrations were calculated in steps of $\Delta r = 100$ meters for the range between 0 and 100 kilometers away from the point of emission, for various emission heights covering the range from 5 meters (traffic emissions) to 200 meters (tall power plant stacks), and for three different annual mean wind speeds representing the three wind speed classes in Germany (2,5 m/s, 3,5 m/s, 4,5 m/s). Fig. 3.4 shows an exemplary radial concentration profile for particles emitted by Diesel vehicles for the default annual mean wind speed of 3,5 m/s for Germany.

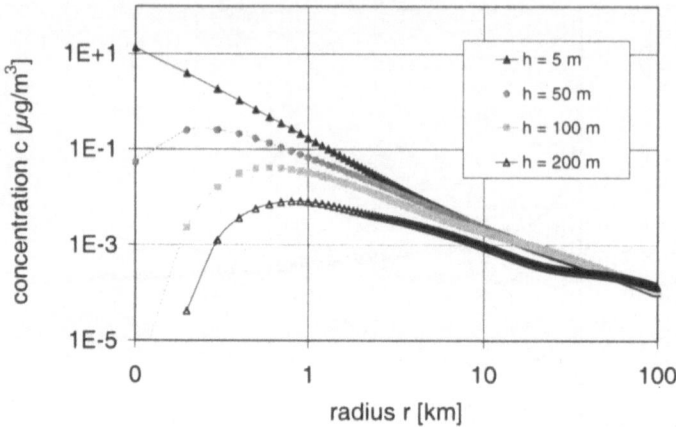

Fig. 3.5. Influence of the effective emission height h on the radial concentration profile for a point source of particulate matter with diameters below 10 μm (PM 10, dry deposition velocity $v_d = 0.25$ cm/s, wet deposition velocity $v_w = 0.92$ cm/s). The emission rate is 1 kg/h.

The sharp decrease of the concentration within the first 10 kilometers results from a combination of vertical dilution (as long as no complete vertical mixing is achieved), horizontal dilution (which essentially introduces a factor proportional to $1/r$), the dry deposition flux[8] (which is assumed to be proportional to $c(r)$) and the increasing transport wind speed. In spite of this sharp decrease, significant contributions to $<I_i>_{C,near}$ also come from the annulus between 10 and 100 kilometers due to its much larger area, compared to the inner circle of radius 10 kilometers. This will be shown in more detail in section 3.6.4. The slower decrease of the pollutant concentration in the range $r > 10$ kilometers is mostly due to horizontal dilution and therefore approximately proportional to $1/r$. Furthermore, it is somewhat influenced by the substance parameters affecting the atmospheric residence time of the pollutants, in particular by the dry deposition velocity v_d, with higher values of v_d leading to lower concentrations.

The pollutant concentration $c(r)$ is essentially inversely proportional to the annual mean wind speed u_m. For a given short-term value u of the wind speed and a given atmospheric stability class, the inverse proportionality is exactly fulfilled as a result of the continuity equation (emission rate $Q \sim 2\pi r \times c(r) \times u$, also see equation (2.20)). Due to the distribution of the short-term wind speeds u around the value u_m and the effect of u on the frequency distribution of the atmospheric stability, the inverse proportionality between $c(r)$ and u_m does not hold exactly, but it was verified to be fulfilled in good approximation.

The effect of the emission height on the pollutant concentration is shown in fig. 3.5 for particulate matter with diameters below 10 μm (PM 10), as they are

---

[8] Despite the higher wet deposition velocity, the dry deposition flux in the vicinity of the source is higher for low emission heights because it is proportional to the ground level concentrations, whereas the wet deposition flux is proportional to the vertically averaged concentration.

emitted from combustion processes such as domestic heating or power plants[9]. Fig. 3.5 shows that the pollutant concentrations in the first few kilometers around the source are very sensitive to the emission height, whereas they become practically independent of the emission height beyond a distance of about 10-20 kilometers due to the complete vertical mixing of the pollutants. The only remaining effect of the emission height at larger distances from the source is related to the amount of dry deposition within the first 10-20 kilometers. This amount is higher for low emission heights, since the dry deposition flux is modeled as being proportional to the ground level concentration of the pollutants (VDI 1992:eq.2). Due to the conservation of mass, higher deposition close to the source leads to lower concentrations further away. However, this effect is not very strong, such that the corresponding intersection of the curves in fig. 3.5 can hardly be discerned.

By combining radial concentration profiles such as the ones shown here with the radial population density distributions for the range of r < 100 kilometers shown above (section 3.3.2), the short-range contributions to the population exposure can be calculated according to equation (3.8). The calculation of the long-range contribution is described in the following section 3.4.2.

3.4.2
Long-Range Transport of Primary Pollutants and Formation of Secondary Pollutants

3.4.2.1
Windrose Trajectory Model

The application of the Gaussian dispersion model described above is limited to distances of up to 100 kilometers away from the emission source (VDI 1992). Concentrations of primary pollutants beyond this distance were calculated using the Windrose Trajectory Model (WTM) which is implemented in the EcoSense software (IER 1998).

The WTM is a receptor-orientated Lagrangian dispersion model. It is based on a model developed to consider the atmospheric transport and chemical transformation of nitrogen and sulfur species (Derwent and Nodop 1986; Derwent et al. 1988). The air parcels considered in the model move along straight line trajectories starting at a travel time of 96 hours (4 days) away from the receptor point. The pollutant concentrations at each receptor result from 24 trajectories, which arrive with receptor-specific wind speeds from different directions in intervals of 15°, and which are weighted with the corresponding frequencies of the wind directions (i.e. the windroses) at the receptor. Besides chemical transformations of atmospheric sulfur and nitrogen species (see below), the model considers dry and wet deposition of pollutants, the latter by using a ‚constant drizzle' approximation. The model uses 1990 meteorological data (windroses, wind speeds, precipitation rates) provided by the Meteorological Synthesizing Centre-West of EMEP at The Norwegian Meteorological Institute (European Commission 1998, Appendix I:6).

---

[9] For the purpose of dispersion modeling, PM 10 was represented by particles with diameters in the range of 1-4 μm, for which v_d = 0,25 cm/s and v_w = 0,92 cm/s, leading to an atmospheric residence time of 22 hours (see table A.1 in the appendix).

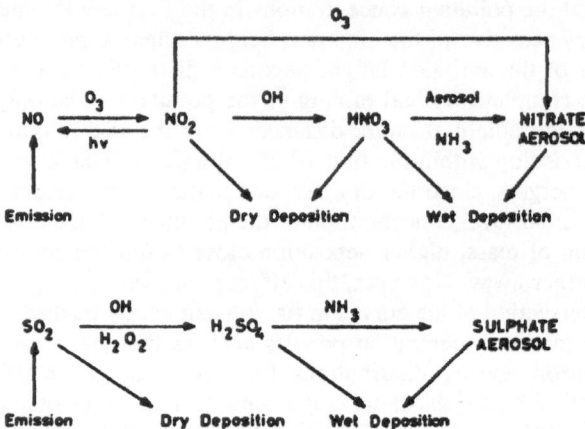

Fig. 3.6. Chemical reaction schemes for the formation of secondary sulfate and nitrate aerosols implemented in the Windrose Trajectory Model (European Commission 1995, vol.4:59). © European Communities. Reproduced by permission of the publisher, the Office for Official publications of the European Communities

In version 2.0 of the EcoSense software, which was used here, the WTM modeling area covers Western and Middle Europe with a North-South extension of 4200 kilometers (between Crete in the Mediterranean Sea at 35°N and the northern end of Scandinavia at 72°N) and a West-East extension of 2700 kilometers (between the Atlantic at 10,5°W and Romania at 30°E). The modeling area is divided into approximately squared grid elements of 10000 km^2 according to the Eurogrid scheme (Bonnefous and Despres 1990).

Pollutant concentrations are therefore calculated with a horizontal resolution of about 100 kilometers. Furthermore, a complete vertical mixing of the pollutants within the constant mixing layer height of 800 meters is assumed to occur *immediately* after their emission, irrespective of the emission height. For these reasons, the model is not suitable for the determination of the short-range contribution to the population exposure, which is strongly sensitive to the emission height (see fig. 3.5) and the variation of the radial population density distribution on a scale of 10 kilometers. For the consideration of the long-range beyond 100 kilometers away from the emission site, the vertical and horizontal resolution of the WTM is sufficient, however.

In addition to dealing with primary pollutants, the Windrose Trajectory Model is also able to calculate concentrations of secondary sulfate and nitrate aerosols formed from emissions of SO_2 and NO_x, respectively[10]. The chemical reaction

---

[10] In fact, the capability of modeling the atmospheric chemistry of secondary aerosols is the main reason for using the model. The long-range population exposures through primary pollutants, on the other hand, can also, with somewhat lower reliability but much smaller effort, be estimated using analytical equations (see section 3.4.2.3).

schemes implemented in the model are shown in fig. 3.6. The secondary sulfate and nitrate aerosols generally represent an important contribution to the overall health impacts of traffic and power plant emissions, as will be discussed in chapter 5. Due to the time it takes before these secondary aerosols are formed, the product $r \times \Delta c(r)$, which determines the contribution to the radial integral in equation (3.8), assumes a maximum at a radius r of some hundred kilometers. A separate consideration of their short-range contribution to the population exposure, including the effects of different emission heights, is therefore not necessary. Instead, the entire population exposure (sum of short and long range) to these pollutants can be calculated in good approximation using the EcoSense software. Since the population exposure to sulfate and nitrate aerosols is not sensitive to the settlement structure class where the emission occurs but only to variations of the population density at a larger geographical scale, country average values will be used here.

3.4.2.2
Pollutant Concentrations and Population Exposures

In combination with a database on population densities contained in the EcoSense software, the WTM was used to determine population exposures in addition to the pollutant concentrations in the area outside a circle with radius 100 kilometers around the emission sites. These long-range contributions to the population exposures were calculated for each of the 41 Eurogrid cells covering the area of Germany. For the example of Diesel particles, fig. 3.7 shows that the assumption of a constant value for $I_{far} \equiv <I_i>_{C,far}$ for all emission sites within one country (Germany in this case) is a reasonable approximation. The standard deviation of the distribution of the values shown in fig. 3.7 is about 30 % of the mean value. If the site of the emission source is known exactly, this variability can be eliminated by using the value for the corresponding Eurogrid cell. This may be particularly required for pollutants with long atmospheric lifetimes, in which case the long-range contribution $<I_i>_{C,far}$ can be larger than the short-range contribution $<I_i>_{C,near}$ (see section 3.6.1).

The long-range contributions to the population exposure to primary pollutants were corrected for the effect that lower emission heights lead to a higher fraction of the pollutants being dry deposited within the circle of radius 100 kilometers around the emission source (for pollutants with $v_d > 0$), such that only a smaller fraction of the pollutants is available for long-range transport beyond 100 kilometers. This correction is in the order of up to 10 % for traffic emissions (h = 5 m) compared to emissions from high stacks (h = 200 m).

For power plant emissions of particulate matter with an atmospheric residence time in the order of one day in Germany, 95 % of the emitted mass are deposited within the modeling area. In the case of SO_2, which is partly transformed into secondary sulfate aerosols, 91% of the sulfur is deposited in the modeling area (European Commission 1995, vol.3:408). Both SO_2 and the secondary sulfates also have an atmospheric residence time in the order of one day. Furthermore, for several thousand kilometers outside of the modeling area, the population density is very low in all directions (with the Atlantic to the West, the Polar Sea to the North, the Mediterranean Sea and Northern Africa to the South and Russia and other states of the former Soviet Union to the East). For substances with atmo-

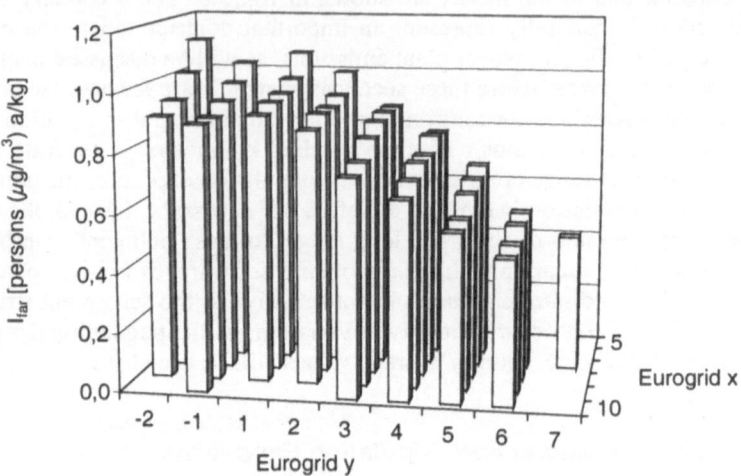

Fig. 3.7. Long-range contribution I_{far} (r > 100 km) to the population exposure per mass of Diesel particles emitted from the 41 Eurogrid elements covering Germany. Eurogrid coordinates x (increasing from West to East) and y (increasing from South to North) according to Bonnefous and Despres (1990)

spheric residence times of up to about one day emitted from Germany, it can be therefore be concluded that there will be no significant contributions to the total population exposure from outside of the modeling domain. In good approximation, this can also be assumed for emissions of such pollutants occurring in countries closer to the borders of the domain. To what extent contributions to the population exposures from outside of the modeling area need to be considered for pollutants with longer atmospheric residence times will be considered in the following section.

3.4.2.3
Analytical Estimates

An analytical expression was used to estimate contributions to the population exposures from outside of the EcoSense modeling area in the case of pollutants with atmospheric lifetimes of more than one day. Assuming a uniform distribution of wind directions and a constant high-altitude transport wind speed u (= 8,6 m/s as an average of the WTM data, IER 1998) as well as an immediate vertical mixing within the constant mixing layer height H (as in the Windrose Trajectory Model), the concentration of a primary pollutant with an atmospheric residence time τ_a around a continuous emission source with emission rate Q only depends on the radial distance r from the source and is given by

$$c(r) = Q \exp(-r/u\tau_a) / (2\pi r H u) \tag{3.14}$$

(Krewitt 1996:32). The corresponding fate factor according to equation (2.13) is given by

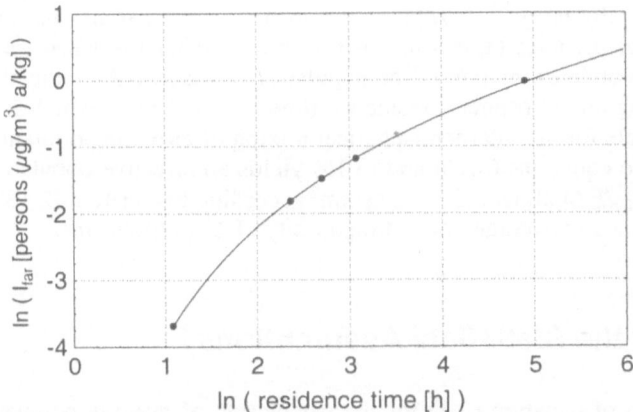

Fig. 3.8. Dependence of the long-range population exposure per mass of emitted particles (I_{far}) of five different diameter ranges in Germany on their atmospheric residence time (as given in table A.1 in the appendix). The interpolation curve is given by equation (3.16).

$$F = \tau_a / H. \tag{3.15}$$

For the population exposure through emission of Diesel particles (residence time about 5 days) in Germany, which was calculated above using the EcoSense software, equation (3.14) was used to estimate the contribution from outside of the modeling area. For this purpose, the outside population density was set equal to the world average (including oceans) of 13 persons/km² (Hofstetter 1998:243), which appears to be a good approximation considering a world map of population densities (Zahn et al. 1996:234-235). Furthermore, the modeling area of 2700 km × 4300 km was replaced by a circle of equal area, i.e. with radius 1922 km. This yields an estimate for the outside contribution to the population exposure per mass of emitted pollutant of 0,16 persons ($\mu g/m^3$) a/kg, which is a small correction to the country average value of the inside contribution for traffic emissions of 1,26 persons ($\mu g/m^3$) a/kg, or to the country average value of 1,06 persons ($\mu g/m^3$) a/kg for emissions at h = 200 meters.

With regards to substances that are not implemented in the EcoSense software, an interpolation formula was derived to determine the long-range population exposure I_{far} per mass of pollutant emitted in Germany as a function of their atmospheric residence times τ_a. For this purpose, a set of five types of particles with different diameters, which determine their dry and wet deposition velocities and hence their atmospheric residence times τ_a, was used. Their substance data and the resulting residence times, which range from 3 hours to 5 days, are shown in table A.1 in the appendix. The interpolation curve is shown in fig. 3.8 and is given by

$$I_{far} \text{ [persons } (\mu g/m^3) \text{ a/kg] } = 0,0205 \ \ln(\tau_a[h])^{2,42}. \tag{3.16}$$

It was applied to determine the long-range population exposures for emissions of carcinogenic substances (benzene, acetaldehyde, formaldehyde, benzo[a]pyrene and 1,3-butadiene) with atmospheric residence times τ_a between 6 hours and 9 days (see table A.1 in the appendix). In the case of benzene ($\tau_a = 9$ days), equation (3.16) was thereby used to *extra*polate to residence times somewhat higher than those of Diesel particles ($\tau_a = 5$ days). For much higher residence times, (3.16) may lead to an underestimation of the population exposures. An indication thereof is that inserting an atmospheric residence time of $\tau_a = 1$ year, which is a characteristic timescale for the interhemispherical mixing of airborne pollutants (Bliefert 1995:106), into equations (3.16) and (3.15) yields an effective population density $\rho_{eff} = I_{tot}/F \approx I_{far}/F$ of about 3,5 persons/km^2 according to equation (2.18), which is lower than the world average population density of 13 persons/km^2.

3.5
Validity of the Statistical Assumptions

The derivation of equation (3.8) for the calculation of average population exposures for the generic spatial classes C rested on two statistical assumptions:

(A1) The variation of meteorological conditions (in particular the wind direction distributions) and the variation of population density with the emission site i are statistically uncorrelated, i.e. $\rho_i(r,\varphi)$ and $\Delta c_i(r,\varphi)$ are uncorrelated with regards to variation of i for each fixed (r,φ).

(A2) The class-averaged population density distribution $<\rho_i(r,\varphi)>_C$ is independent of the angle φ, i.e., on average, there is no preferred direction for the spatial variation of ρ.

While both assumptions appear to be intuitively plausible, they were also tested in a quantitative way for the settlement structure classes introduced in section 3.3.1. Since the two assumptions were used in the calculation of the short-range contribution $<I_i>_{C,near}$ to the population exposure, the range of 0 to 100 kilometers for the variable r and the range of 0 to 2π for the variable φ were considered. As for the calculation of the radial population density distributions, the r-range was discretized into ten equal intervals in steps of 10 kilometers, while the φ-range was discretized into 12 angular sectors. For each municipality in Germany (i = 1.. 14615), the two-dimensional population density distribution $\rho_i(r,\varphi)$ was calculated in a way analogous to the calculation of the radial population density distributions $\rho_i(r)$.

A strict calculation of $\Delta c_i(r,\varphi)$ would require the availability of locally specific meteorological data, in particular of locally specific ‚windroses' $w_i(\varphi)$, i.e. frequency distributions of the wind directions (normalized to 1). For reasons discussed above (section 3.1), these are not easily available. As an approximation for the test of (A1), windroses referring to an altitude of about 600 meters above the ground were used. These are less locally variable as they are not strongly influenced by topographical features. The high altitude windroses used here were taken from the data incorporated in the EcoSense software tool (IER 1998). They are based on meteorological data for the year 1990 provided by the Meteorological Synthesizing Centre-West of EMEP at The Norwegian Meteorological Institute, and given with a spatial resolution of 100 km × 100 km.

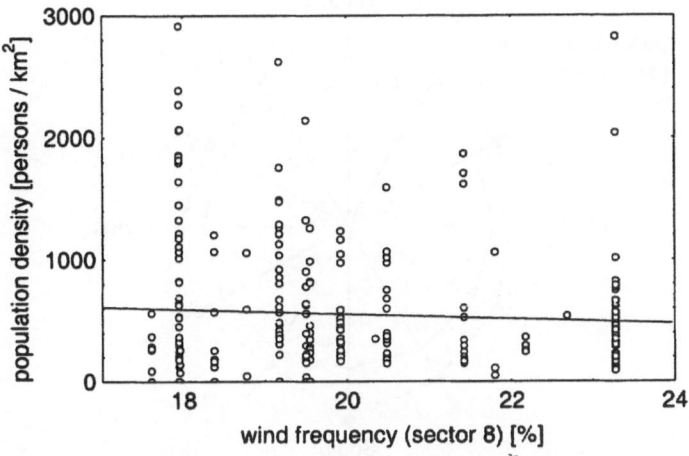

Fig. 3.9. Test of the statistical assumption (A1) for the settlement structure class (I,2,a) (264 municipalities in Germany close to major cities in agglomerated regions), showing the population density $\rho_i(r,\varphi)$ for the radial interval [10 km, 20 km[and the direction 2 o'clock versus the wind frequency w_i for the predominant wind direction (8 o'clock) for all sites i within the class. The linear regression curve is almost horizontal (regression coefficient $r = -0,06$), indicating that there is practically no correlation. Hence, (A1) is fulfilled in very good approximation in this case.

These windroses $w_i(\varphi)$, of which there are about 40 for the various parts of Germany, can be used to calculate the incremental concentration $\Delta c_i(r,\varphi)$ with the Gaussian dispersion model that was also used to calculate the class-averaged radial concentration profiles $<\Delta c_i(r)>_C$ (see section 3.4). However, instead of running the model 40 times, the angular dependence of the concentration was represented in a simplified way for the purpose of testing (A1) by using the factorization

$$\Delta c_i(r,\varphi) = <\Delta c_i(r)>_C \times w_i(\varphi-\pi). \tag{3.17}$$

The phase shift of π in the argument of w_i is thereby due to the fact that the concentration in one particular direction is related to the frequency of the wind coming from the opposite direction. The class C was specified by using an annual mean wind speed of 3,5 m/s, which is the default value for Germany. The factorization (3.17) neglects the fact that, due to the lateral dispersion of air pollutants, the angular distribution of concentrations can be expected to be smoother than the angular distribution of the wind directions. This would imply an overestimation of the angular variation of the concentration. On the other hand, the use of the high-altitude windroses alone would lead to an underestimation of this variation. Both approximations taken together can therefore be considered as a reasonable first approximation of the actual angular distribution of the pollutant concentration.

Having calculated $\Delta c_i(r,\varphi)$ and $\rho_i(r,\varphi)$ for each municipality in Germany in this way, the assumptions (A1) and (A2) were tested for the settlement structure classes introduced in section 3.3. Assumption (A1) was tested by calculating, for

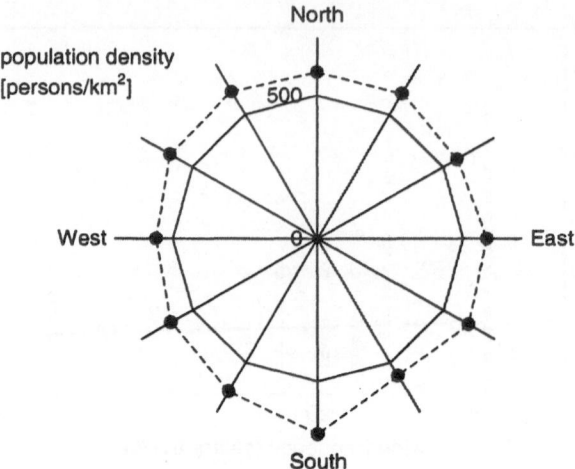

Fig. 3.10. Test of the statistical assumption (A2) for the settlement structure class $C = (I,2,a)$ (264 municipalities in Germany close to major cities in agglomerated regions), showing the population density $<\rho_i(r,\varphi)>_C$ for the radial interval $[10\ km,\ 20\ km[$ and φ discretized into 12 angular sectors. The standard deviation of the distribution of $<\rho_i(r,\varphi)>_C$ is 6 % of the mean value, indicating that $<\rho_i(r,\varphi)>_C$ is practically independent of φ. Hence, (A2) is fulfilled in very good approximation in this case.

the sites i within each of the 17 classes at the municipality level, the Pearson product-moment correlation coefficients r between the values $\rho_i(r,\varphi)$ and $w_i(\varphi-\pi)$ (being the only part of $\Delta c_i(r,\varphi)$ varying with i according to (3.17)) for each of the 120 combinations of the 10 radial intervals and the 12 angular sectors. The values of the correlation coefficients r turned out to be consistently low, typically in the absolute range of 0,1 to 0,3 with positive and negative values occurring with practically equal frequency, indicating that no strong correlations were found. Therefore, assumption (A1) was found to be fulfilled in very good approximation.

For the settlement structure class (I,2,a), which contains 264 municipalities that are close to major cities in agglomerated regions, one of the 120 investigated correlations, namely for the radial interval of $[10\ km,\ 20\ km[$ and the angular sector corresponding to the average predominant wind direction is shown in fig. 3.9. The relevance of this particular example is that the interval of $[10\ km,\ 20\ km[$ can be expected to contribute significantly to $<I_i>_{C,near}$ for this class due to its high population density, which corresponds to the nearby location of a major city (see fig.3.2(a) for the more general class (I,2)). Therefore, the values of $I_{i,near}$ for the sites i within this class can be expected to be quite sensitive to the direction where the nearby major city is situated relative to the predominant wind direction. The relevance of assumption (A1) being fulfilled in this case is that, nevertheless, the class average $<I_i>_{C,near}$ is insensitive to this directional effect.

Assumption (A2) was tested by calculating $<\rho_i(r,\varphi)>_C$ for each of the 17 combined NUTS level 5 classes for each of the 120 combinations of the 10 radial intervals and the 12 angular sectors. The φ-independence of this expression was also found to be fulfilled in very good agreement. For a fixed radial interval, the distribution of the values of $<\rho_i(r,\phi)>_C$ for the 12 ϕ-intervals was typically very narrow (the ideal case being a uniform distribution), with standard deviations typically in the range of 10% to 30% of the mean value. An example is shown in fig. 3.10, again for the settlement structure class (I,2,a) and the radial interval [10 km, 20 km[.

It should be noted that the statistical assumptions (A1) and (A2) are fulfilled in good approximation also for those settlement structure classes with the smallest number of elements, namely the classes (I,1,a) (n = 12), (I,1,b) (n = 41) and (II,1) (n = 29), all of which consist of central cities. Furthermore, the size of the population exposure for emission at the central cities constituting these classes is not very sensitive to the angular distribution of the population density around them, since the largest share of the exposure comes from the cities themselves (see sections 3.6.3 and 3.6.4).

3.6
Results and Discussion[11]

3.6.1
Traffic Emissions

The spatial differentiation of the impacts due to the influence of the population density is shown in fig. 3.11 for the example of traffic emissions of acetaldehyde, which has a relatively short atmospheric residence time of 9 hours, mainly due to reaction with hydroxyl radicals (Seinfeld and Pandis 1998:110, 288). The population exposure per mass of the emission (denoted as I_C as a shorthand for $<I_i>_C$) is shown on the y-axis as a measure of the impacts, since the effect factor E in equation (3.1) is the same in all cases. The short-range contribution to I_C was calculated for the default annual mean wind speed of 3,5 m/s for Germany. Since the focus is on the degree of spatial differentiation rather than on the absolute values of the impacts, I_C is normalized to the reference situation of a constant radial population density distribution equal to the country average value of 230 persons/km² for Germany (BBR 1998a). This German average value of the population exposure per mass of emitted acetaldehyde is I_D = 0,62 persons (μg/m³) a/kg.

Fig. 3.11 shows the relative impacts for five out of the nine combined settlement structure classes at the district (NUTS 3) level, with the values for the other four classes being close to the reference value. The impact is highest for the class (I,1) of central cities within agglomerated regions, and lowest for the class (III,5) of low-density rural districts in rural regions. The overall spread between the two extreme classes at the district level in Germany is by a factor 5 in this example. The impacts for the class of highly densified districts which are close to major cities in agglomerated regions (I,2) and for ‚normal‘ cities (i.e. central cities within urbanized regions (II,1)) are of a similar size significantly above the country average value. The impacts for the remaining five settlement structure classes are close to the country average value. Fig. 3.11 includes one of these classes,

---

[11] A summary of the calculated population exposures can be found in table A.2 in the appendix.

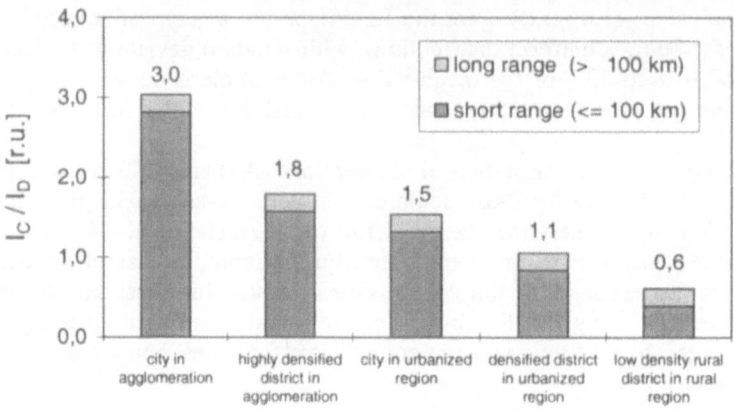

Fig. 3.11. Influence of the settlement structure class C on the population exposure per emitted mass I_C through airborne traffic emissions of acetaldehyde (atmospheric residence time 9 hours). The population exposures are normalized to a country average value of $I_D = 0{,}62$ persons ($\mu g/m^3$) a/kg corresponding to the average population density of 230 persons/km² for Germany. The population exposures are shown for five out of the nine settlement structure classes at the district level. The values for the remaining four classes are close to the country average value. A default annual mean wind speed of 3,5 m/s for Germany was used.

namely the class of densified districts in urbanized regions, as a typical example of an ‚average' situation.

It was found that a further subdivision of these nine combined district (NUTS level 3) classes into the 17 combined NUTS level 5 classes does not, in general, lead to further significant differentiations of the impacts relative to the intra-class variability of the impacts discussed in section 3.6.3. The only exception to this is the distinction between cities with more than 500.000 inhabitants (class (I,1,a), n = 12) and cities with less than 500.000 inhabitants (class (I,1,b), n = 46) within the class (I,1) of central cities within agglomerated regions. In the example of acetaldehyde, the respective impacts are 4,0 times the country average I_D for (I,1,a) and 2,6×I_D for (I,1,b) instead of 3,0×I_D for (I,1). This differentiation increases the spread between the class with the highest impacts (I,1,a) and the class with the lowest impact (III,5) to a factor 6,7.

Average impacts were, on the other hand, also calculated for the larger region classes I, II and III (agglomerated, urbanized and rural regions). The averages for class I and class III turned out to differ only by a factor of 2, indicating that the region classes alone are too rough to represent the degree of spatial differentiation of the impacts. Instead, their combination with classes at the district level is necessary. Nevertheless, some of the nine combined settlement structure classes at the district level can be merged because they show average impact values that are not significantly different from each other. This will be further discussed in section 3.6.3 in the light of the remaining variability of the impacts within the classes.

Fig. 3.11 shows the values of the population exposure as the sum of the short-range contribution (r ≤ 100 km) and the long-range contribution (r > 100 km). Due to the short atmospheric residence time of acetaldehyde (9 hours), the impacts are

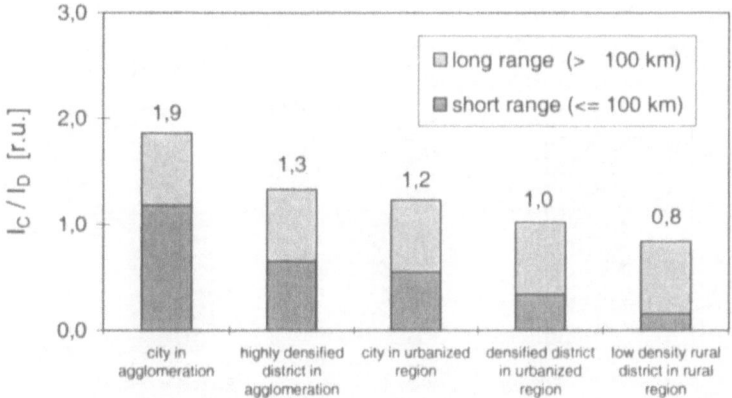

Fig. 3.12. Influence of the settlement structure class C on the population exposure per emitted mass I_C through airborne traffic emissions of Diesel particles (atmospheric residence time 5 days). The population exposures are normalized to the country average value of $I_D = 1,41$ persons ($\mu g/m^3$) a/kg corresponding to the average population density of 230 persons/km^2 for Germany. The population exposures are shown for five out of the nine settlement structure classes at the district level. The values for the remaining four classes are close to the reference value. A default annual mean wind speed of 3,5 m/s for Germany was used.

dominated by the short-range contribution, but even in this case, the long-range contribution cannot be entirely disregarded. In the case of substances with longer atmospheric residence times, the long-range contribution becomes more significant. This is shown in fig. 3.12 for the case of Diesel particles with an atmospheric residence time of about 5 days. The impacts are again normalized to the reference situation defined above. The country average value for traffic emissions of Diesel particles is $I_D = 1,41$ persons ($\mu g/m^3$) a/kg. Since the long-range contribution is assumed to be the same for all settlement structure classes, its higher contribution to the overall impacts leads to a reduction of the spread between the extreme values from a factor 5 in the example of acetaldehyde to a factor 2,2 for Diesel particles. At the same time, the effect of longer atmospheric residence times is to increase the absolute values of the population exposures for all settlement structure classes.

Since the pollutant concentration in the short range ($r < 100$ km) is approximately inversely proportional to the annual mean wind speed, this also applies to the short-range contribution to the population exposure. From section 3.4.1.2, it may be recalled that the annual mean wind speed varies between 2,5 m/s and 4,5 m/s in most parts of Germany, if considered with a spatial resolution corresponding to the linear extension of the short range, i.e. 100 kilometers. Therefore, the consideration of the annual mean wind speed introduces a spread of the short-range population exposures in the order of a factor 4,5/2,5 = 1,8. Detailed calculations yield a somewhat smaller factor of 1,5. Higher windspeeds are thereby associated with lower population exposures. Fig. 3.13 shows how the country average value of the population exposure is affected by changing the annual mean wind speed from the default value of 3,5 m/s to either 2,5 m/s or 4,5 m/s in the case of

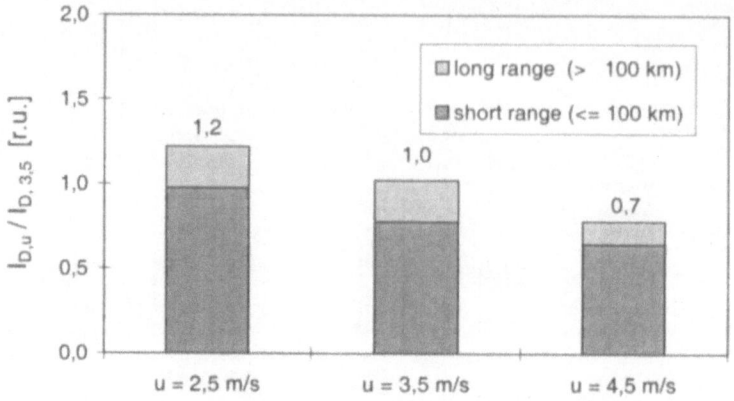

Fig. 3.13. Influence of annual mean wind speed u on the population exposure per emitted mass $I_{D,u}$ through airborne traffic emissions of acetaldehyde (atmospheric residence time 9 hours). $I_{D,u}$ refers to a constant radial population density distribution at the country-wide average value of 230 persons/km² for Germany and is normalized to the annual mean wind speed of 3,5 m/s ($I_{D, 3.5} = 0,62$ persons ($\mu g/m^3$) a/kg).

an emission of acetaldehyde. For reasons of consistency between the spatial variation of the annual mean wind speed according to fig. 3.3 and the spatial variation of the long-range contribution to the population exposure (see section 3.4.2.2), the latter also varies in fig. 3.13.

In cases where the short-range contribution dominates the total population exposure (i.e. for substances with short atmospheric residence times, and for large cities in agglomerations even in the case of longer residence times), the annual mean wind speed therefore affects the total impact values to some extent. Compared to the differentiation of the impacts due to the population density, the spread of the impacts introduced by the variation of the wind speed is significantly smaller, however (e.g. a factor 1,7 due to the wind speed as opposed to a factor of 5 due to the settlement structure in the above example of acetaldehyde). Whether or not it is worthwhile to consider the influence of the annual mean wind speed can therefore be decided case by case, bearing in mind that this requires more detailed knowledge of the location of the emission source (see fig. 3.3) than a differentiation on the basis of the settlement structure classes alone.

3.6.2
Influence of the Emission Height

The same calculations as for traffic emissions were also carried out for other emission heights up to 200 meters. The long range contribution is essentially independent of the emission height, except for small changes due to the correction for the amount of dry deposition within the short range. Higher emission heights are thereby associated with a slightly higher long-range contribution to the population exposures. The main effect, however, is the decrease of the short-range contribu-

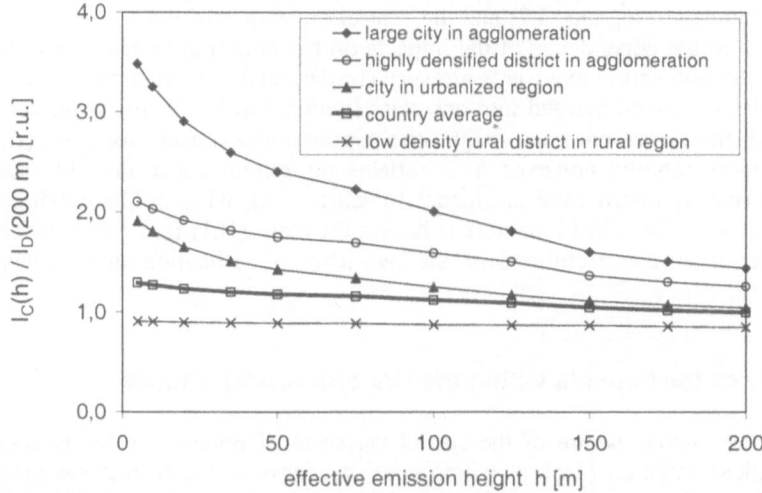

Fig. 3.14. Influence of the effective emission height h and the settlement structure class C on the population exposure per emitted mass $I_C(h)$ through airborne emissions of PM 10 ($v_d = 0{,}25$ cm/s, $v_w = 0{,}92$ cm/s, $\tau_a = 22$ hours). $I_C(h)$ is normalized to the country average value $I_D(200\,m)$ $= 0{,}49$ persons $(\mu g/m^3)$ a/kg. It is shown for four out of the nine settlement structure classes at the district level and the country average situation, which approximately represents the remaining five classes. A default annual mean wind speed of 3,5 m/s for Germany was used.

tion with increasing emission height due to the decrease of the ground-level concentrations within the first 10-20 kilometers around the emission source (see fig. 3.5). This decrease of the short-range contribution is strongest for those settlement structure classes with a high population density within the first 20 kilometers (i.e. (I,1), (I,2) and (II,1)), whereas the impacts for the other classes are less strongly affected by the emission height.

This is shown in fig. 3.14 for four out of the nine settlement structure classes together with the country average situation, which approximately represents the remaining five classes. The example substance is particulate matter with diameters below 10 μm (PM 10)[12], which is emitted, e.g., from the combustion of oil or coal in domestic heatings and power plants. The country average value for the normalization of the population exposure was calculated for an emission height of h = 200 meters ($I_{D,\,200} = 0{,}49$ persons $(\mu g/m^3)$ a/kg). A default value of 3,5 m/s for the annual mean wind speed for Germany was used throughout.

Fig. 3.14 shows that the degree of spatial differentiation of the impacts decreases with increasing emission height. While the spread between the extreme classes (I,1) and (III,5) is by a factor of 3,8 in the case of traffic emissions (h = 5 m), it only amounts to a factor 1,7 for emissions from high stacks

---

[12] For the purpose of dispersion modeling, PM 10 was represented by particles with diameters in the range of 1-4 μm, for which $v_d = 0{,}25$ cm/s and $v_w = 0{,}92$ cm/s, leading to an atmospheric residence time of 22 hours (see table A.1 in the appendix).

(h = 200 m). Furthermore, it is interesting to compare the highly densified districts within agglomerated regions (I,2) and the central cities in urbanized regions (II,1) with regards to the dependence of the impacts on the emission height. For traffic emissions, the population exposures are very similar, such that the two settlement structure classes can be merged into one class (which will be discussed further in 3.6.3). With increasing emission height, the populations exposure for class (II,1) decreases more rapidly, however. This reflects the circumstance that the radial population density distribution is broader for class (I,2), whereas the population density within the first 10 kilometers is higher for class (II,1) (see fig. 3.2). This again underscores the usefulness of the two-level classification of population densities.

3.6.3
Variability of the Impacts within the Generic Spatial Classes

Given the continuous nature of the spatial variation of population densities and meteorological variables (such as windspeed), the introduction of discrete spatial classes is inevitably associated with a remaining variability of the impacts within the classes. While these intra-class variabilities cannot be avoided, a suitable classification is one that keeps the overlap between the classes small while providing a significant variation of the impacts between the classes[13].

While the exact variability of the impacts within each class can only be known if all individual impact values are known exactly (which is not practically feasible for the large number of sites to be considered), estimates of the intra-class variability of impacts can nevertheless be provided. With regards to the short-range contribution to the population exposure, such estimates were calculated in two ways.

On the one hand, estimates for the actual population exposures per mass I_i associated with emissions at the individual sites i within the settlement structure classes were derived by using the individual radial population density distributions $\rho_i(r)$ in equation (3.8) instead of their class average $<\rho_i(r)>_C$. These estimates will be denoted as I_{1i} in the following. For each settlement structure class, the standard deviation ΔI_1 of their distribution was used as an indication of the intra-class variability of the actual values of I_i. To be more precise, ΔI_1 captures the variability of the population exposures due to *radial* variations of the population density around the sites[14].

A second component of the intra-class variability of the population exposures is associated with *angular* correlations between the population density and the wind direction for individual sites (which do not influence the class average according to equation (3.4)). Estimates I_{2i} of the actual values I_i which take both the radial and the angular dimension into account were calculated according to equation (3.2) by using the approximations for the pollutant concentrations $\Delta c_i(r,\varphi)$ derived on the basis of high-altitude windroses and the two-dimensional population densi-

---

[13] In contrast, a random classification of sites can be expected to yield the same average impact value and the same variability of impacts for each class.

[14] The annual mean wind speed is kept constant here, since the largest share of the variability is due to the population density.

ties $\rho_i(r,\varphi)$, as they were introduced in section 3.5. Since the determination of $\Delta c_i(r,\varphi)$ in this way was shown to be a good compromise between over- and underestimating the variability of the actual concentrations, this also applies to estimating the variability of I_i on the basis of I_{2i}[15].

Due to the inclusion of both the radial and the angular variability, the standard deviation ΔI_2 is generally larger than ΔI_1. For traffic emissions (h = 5 m), ΔI_1 is typically in the order of 35 % of the mean value of the population exposure for each of the nine settlement structure classes. The standard deviation ΔI_2 of the distribution of the I_{2i} values is in the order of 45-50 %. This means that about 50 to 60 % of the *variance* of I_i (i.e. the sum of the *squared* deviations from the class average) is due to the radial variability and 40 to 50 % due to the angular variability[16].

However, for the settlement structure classes (I,1) and (II,1), both standard deviations (and in fact the individual values of I_{1i} and I_{2i}) were practically the same, meaning that the overall variability can be explained by the radial variability alone. This is plausible considering that these classes represent central cities where most of the population exposure comes from within the first 10 kilometers around the source (see 3.6.4) and is therefore not sensitive to any angular effects, given the radial resolution of 10 kilometers. Finally, it should be noted that, with increasing emission height, the intra-class variability decreases in the same way as the differences between the class-averages.

The variability of the long-range contribution to the population exposure was investigated by running the Windrose Trajectory model for identical emissions occurring within each of the 41 Eurogrid cells covering the area of Germany. The long-range contributions generally decrease from the southwest to the northeast of Germany due to variations of the population density in Europe on the scale of several hundred kilometers (see fig. 3.7). The standard deviation of the corresponding distribution of long-range population exposures is typically in the order of 30 % of the mean value. Therefore, the simplification of using a constant value of the long-range contribution for all emission sites in Germany is of the same quality as the simplification of using the class-average short-range contribution for all sites within one settlement structure class. In that sense, both simplifying steps are consistent. In cases where the short-range contribution dominates the total impact, this also applies to the respective variabilities, and vice versa for the long-range contribution. In both cases, the overall variability is in the order or 40 % of the class-average. An example for the first situation is presented in fig. 3.15, which shows the spatial differentiation of the impacts of traffic emissions of acetaldehyde and is very similar to fig. 3.11, except that error bars indicating the intra-class variabilities of the impacts have been added and all nine settlement structure classes are shown.

In the light of the intra-class variabilities for traffic emissions shown in fig. 3.15, the impacts for the classes (I,2) and (II,1) are not significantly different. Both

---

[15] In fact, the values of I_{2i} can also be used for a more site-specific impact assessment for emission sites with exactly known location.

[16] The class average population exposures were also calculated on the basis of I_{2i}. It turned out that $\langle I_{2i} \rangle_C = \langle I_{1i} \rangle_C$ for all settlement structure classes, which provides a further verification of the statistical arguments introduced in section 3.2.

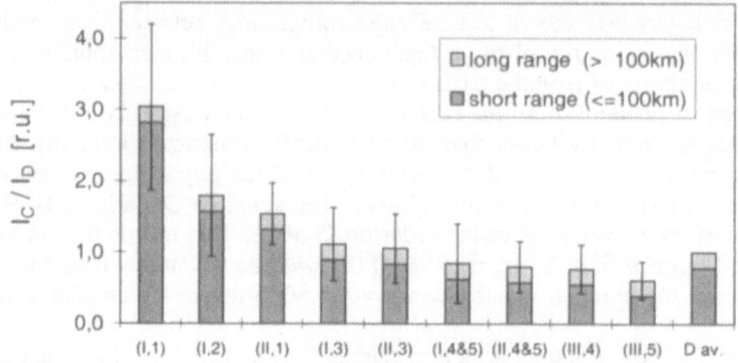

Fig. 3.15. Estimated variabilities of the population exposures per emitted mass I_C through airborne traffic emissions of acetaldehyde within the nine combined settlement structure classes C at the district level (notation as defined in section 3.3.1, *D. av.* country average). The country average value is $I_D = 0{,}62$ persons $(\mu g/m^3)$ a/kg. The default annual mean wind speed of 3,5 m/s for Germany was used.

settlement structure classes can therefore be merged into one generic spatial class for traffic emissions (but not for higher emission heights, as was discussed above). The same applies to the group consisting of the classes (I,3), (II,3), (I,4&5), (II,4&5), (III,4) and (III,5) for which the impacts form a continuum around the country average value for all emission heights. Depending on the emission height, this leaves three or four distinct generic spatial classes to be discerned on the basis of the nine settlement structure classes at the district level.

3.6.4
Relevance of Various Distance Ranges and of Street Canyons

In addition to splitting the population exposures into a long-range contribution ($r > 100$ km) and a short-range contribution ($r \leq 100$ km), the contributions from various distance ranges to the latter were examined in more detail by calculating the function

$$I_C(R) = \int_0^R <\rho_i(r)>_C <\Delta c_i(r)>_C \, 2\pi r \, dr \tag{3.18}$$

in the range $R = 0..100$ km ($I_C(100 \text{ km}) = I_{C,near}$) for the various settlement structure classes. Fig. 3.16 shows this function for the example of traffic emissions of acetaldehyde and a selection of settlement structure classes. In the case of the central cities (classes (I,1) and (II,1)), about 65 % of $I_{C,near}$ come from the first 10 kilometers around the emission source. The other extreme in this regard are the low density rural districts in rural regions (III,5), where the fraction $I_C(10 \text{ km})/I_{C,near}$ only amounts to 21%.

It may be recalled from section 3.4.1.1 that the Gaussian dispersion model used to derive the population exposures assumes a flat terrain around the emission source and the absence of other obstacles. However, traffic emissions within cities

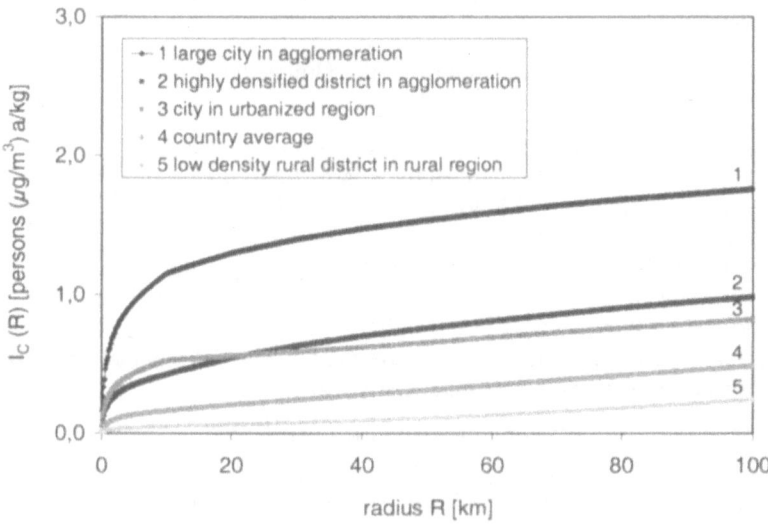

Fig. 3.16. Cumulative contribution $I_C(R)$ to the short-range population exposure per emitted mass $I_{C,near}$ for the distances R between 0 and 100 kilometers away from the emission source, as defined in equation (3.15), for selected settlement structure classes C for the example of traffic emissions of acetaldehyde.

frequently occur in street canyons, which may considerably alter the immission concentrations in the vicinity of the vehicles up to distances of some ten meters. Whether this circumstance has a significant influence on the population exposures can be discussed on the basis of the function $I_C(R)$. Fig. 3.16 shows that, for the case of a flat terrain, the range of a few ten meters around the source, which is the typical width of a street canyon, only contributes negligibly to the overall population exposure. In order for street canyons to make a difference, they would need to influence the pollutant concentrations by several orders of magnitude.

Many models exist to calculate pollutant concentrations in street canyons, ranging from very simple analytical models that calculate annual mean values of concentrations to complex computer models providing high spatial and temporal resolution. The latter are typically designed for regulatory purposes in order to examine possible exceedances of limit values of immission concentrations at particular points in specific streets. In Germany, for example, such models can be used instead of measurements in order to test the compliance of immission concentrations with the 23rd Amendment to the German Air Pollution Control Act (23. Bundes-Immissions-Schutz-Verordnung, see, e.g., Seidler 1994).

In the context of the present study, however, it is the integral of the concentration of a pollutant over a large area (circle of radius 100 km) rather than the concentrations at particular points in space that is of interest. It is therefore sufficient to obtain an idea of the order of magnitude by which immission concentrations are changed in street canyons compared to an unobstructed dispersion of pollutants. This can be achieved by using simple street canyon models, which are also often referred to as „screening models" in regulatory contexts. Two analytical screening models were used here:

1. The SRI (Stanford Research Institute) model (Johnson et al. 1973; Dabberdt et al. 1973)

The SRI model is based on an evaluation of carbon monoxide (CO) immission concentrations at seven measurement stations in a street canyon in San Jose, California. Its ratio of height to width was in the order of 1. For the case of the wind blowing perpendicular to the street, the model distinguishes between the lee side and the luv side of the canyon. Lee and luv thereby refer to the wind direction at the bottom of the canyon, which is opposite to the rooftop wind direction due to the formation of a vortex. For each case, the incremental pollutant concentration caused by the emissions in the street (to be distinguished from the background concentration) is represented by a simple analytical formula:

$$\Delta c_{luv}(z) / q_l = K (H-z) / [(u+u_0) H W]$$ (3.19)

$$\Delta c_{lee}(x,z) / q_l = K / \{(u+u_0) [(x^2+z^2)^{1/2} + L]\}$$ (3.20)

with

q_l	emission rate of the street [mass time^{-1} length^{-1}]
H	height of the street canyon
W	width of the street canyon
x	horizontal distance from lane
z	vertical distance from street level
u	annual mean wind speed
u_0	= 0,5 m/s (empirically determined)
K	= 7 (empirically determined)
L	= 2 m (empirically determined).

The parameters K, u_0 and L are empirical inputs into the model. A more recent regression analysis of immission measurements in Zurich, Switzerland yielded somewhat different values for these parameters, which were also used here: K = 17,5 for the luv side and K = 13, L = 5,94 m for the lee side (Pelli 1985:350).

2. The Dutch CAR (Calculation of Air Pollution from Road Traffic) model (Eerens et al. 1993, den Boeft et al. 1996)

This model is a parameterization of a more complex model which is based on measurements, wind tunnel experiments and theoretical considerations. It distinguishes five street types according to the degree of obstruction of the dispersion of pollutants. The street types range from situations of unobstructed dispersion to narrow street canyons. In each case, the incremental concentration caused by the emissions in the street is written as

$$\Delta c(s) / q_l = \Phi(s) F_t (u_{standard}/u_{regional})$$ (3.21)

with

q_l	emission rate of the street [mass time^{-1} length^{-1}]
s	distance from the middle of the street [m]

$\Phi(s)$ a quadratic function of s depending on the street type (see below) [s/m^2]

F_t factor that accounts for the presence of trees in the street ($F_t = 1$ in the absence of trees, $F_t = 1,25$ or $1,5$ with trees)

$u_{standard}$ standard value of the annual mean wind speed in the Netherlands, measured at Schipol airport ($u_{standard} = 5$ m/s, Troen and Petersen 1989:387)

$u_{regional}$ annual mean wind speed in the area under consideration.

The function $\Phi(s)$ is defined as

$$\Phi(s) = as^2 - bs + c \tag{3.22}$$

with parameters a, b, c > 0 listed in table 3.5 for the different street types. In order to compare the street canyon screening models SRI and CAR with the Gaussian model used here, a few simplifications can be made for the latter: Since only distances from the source of some ten meters are considered, the effects of pollutant deposition and reflection of the plume at the top of the mixing layer can be neglected. In this case, the Gaussian pollutant concentration profile downwind of a street with emission strength q_l, emission height $h = 0$ and the wind blowing perpendicular to the street is

$$\Delta c(x,z) / q_l = (2/\pi)^{1/2} \, 1 / [u \, \sigma_z(x)] \, \exp[- z^2 / (2\sigma_z(x)^2)] \tag{3.23}$$

(Seinfeld and Pandis 1998:944-945, equations 18.70 and 18.76 with $p = 0$, $n = 0$, $\alpha = 2$, $\nu = 1/2$). Using the parameterization

$$\sigma_z(x) \, [m] = 0,215 \, x[m]^{0.885} \tag{3.24}$$

for the atmospheric stability class III/1, which is dominant during daytime when most traffic emissions occur (VDI 1992), and the default value of $u = 3,5$ m/s for the annual mean wind speed in Germany, this reduces to

$$\Delta c(x,z = 0) \, [sm^{-2}] = 1,06 / x[m]^{0.885}. \tag{3.25}$$

Fig. 3.17 shows the concentration profiles in the vicinity of a street for this Gaussian dispersion model, the SRI model (with parameters modified according to Pelli 1985) and the two street types of the CAR model with the highest and lowest concentration values, respectively. The results for the Gaussian model and the CAR model for street type 1, which both describe an unobstructed dispersion of pollutants, are in fairly good agreement for distances of more than 5 meters away from the source. Closer to the source, the CAR model for street type 1 predicts much lower concentrations. This may to some extent be due to traffic induced

Table 3.5. Parameters of the CAR model depending on the street type (Eerens et al. 1993:391)

street type	description	a	b	c
1	open terrain	0,75 E-4	0,70 E-2	0,17
2	all other than 1,3,4	3,10 E-4	1,82 E-2	0,33
3a	wide street canyon	3,25 E-4	2,05 E-2	0,39
3b	narrow street canyon	4,88 E-4	3,08 E-2	0,59
4	one-sided street canyon	5,00 E-4	3,16 E-2	0,57

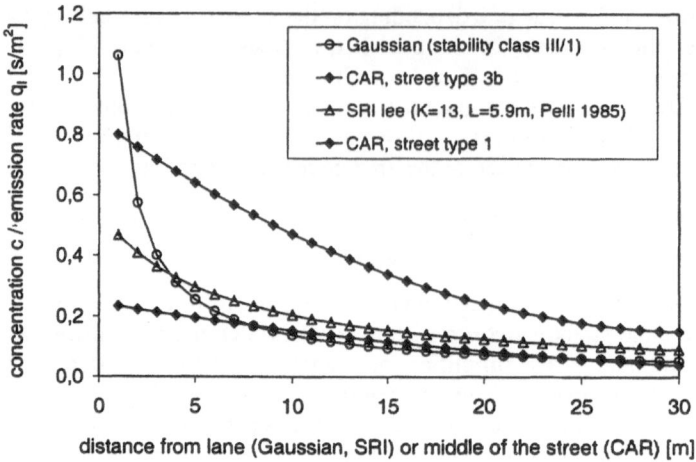

Fig. 3.17. Comparison of concentration profiles for a Gaussian line source model for unobstructed dispersion, the SRI screening model for street canyons (with parameters according to Pelli 1985) and the CAR screening model (*street type 1*: unobstructed dispersion, *street type 3b*: narrow street canyon)

turbulence, which reduces concentrations and is neglected in the Gaussian model. Furthermore, the concentrations in the CAR model represent the combined effect of emissions from more than one street lane, whereas the Gaussian model describes the idealized situation of a one-dimensional line source with no lateral extension.

The effect of the obstruction of the pollutant dispersion by a street canyon is to increase the pollutant concentrations within the canyon. According to fig. 3.17, this effect is strongest for street type 3b of the CAR model, which represents a narrow street canyon. The concentrations predicted by the CAR model for this street type are up to about 3,5 times as high as those predicted by the Gaussian model. Since the CAR model for street type 3b generally predicts concentrations which are higher than the values produced by other models (SRI and IMMIS-Luft) as well as the measured values (Bächlin et al. 1995:45), its results can be considered as an upper limit for the actual concentrations in street canyons. This is supported by a study that showed measured concentrations for benzene in street canyons at 14 inner-city stations in Germany to be only about a factor 2 higher than those predicted by a Gaussian dispersion model (Friedrich and Schierbaum 1997).

While pollutant concentrations may be about 2-3 times as high as those predicted by a Gaussian model within street canyons, this nevertheless has a negligible effect on the integral of the pollutant concentration over the range of 100 kilometers, since the width of the canyons is only in the order of some ten meters.

It may be argued that street canyons also affect the pollutant concentrations on the outside due to shifting the emission height from close to the ground for vehicles to the height of the rooftops, which may be some ten meters. Considering the dependence of the population exposure on the emission height (section 3.6.2), this might have some effect on the overall impact. However, most street canyons are not impermeable since pollutants also disperse between buildings and along other streets at crossroads. Furthermore, the exposure of people living a few storeys above ground would then also have to be considered. It appears to be qualitatively plausible that the overall effect of the altered emission height would be significantly diminished. This effect shall therefore be neglected in the following.

While considering the effects of street canyons is important if one is interested in pollutant concentrations at particular points within the canyons (e.g. in a regulatory context), it can be concluded from the above considerations that their effect on the population exposure (i.e. the area-integrated product of pollutant concentration and population density) over a range of 100 kilometers is negligible.

3.6.5
Secondary Aerosols

The country average population exposures to secondary aerosols and their standard deviations were determined from 41 runs of the Windrose Trajectory Model for the Eurogrid squares covering the area of Germany as (0.14 ± 0.03) persons $(\mu g/m^3)$ a/(kg of emitted SO_2) for sulfate aerosols and as (0.24 ± 0.10) persons $(\mu g/m^3)$ a/(kg of emitted NO_2) for nitrate aerosols. Due to the time required for the formation of the secondary pollutants, the short-range contribution (r < 100 km) to these population exposures is typically less than 10 %. The variability is due variations of the population densities in Europe on the scale of several hundred kilometers. The population exposures therefore generally decrease from the southwest to the northeast of Germany.

3.6.6
Default Values for Other Countries

The definition of generic spatial classes to capture the spatial differentiation of health impacts of airborne pollutants with linear exposure response functions was demonstrated for the case of Germany. Based on a similar classification of settlement structures in other Western European countries (Schmidt-Seiwert 1997) or extensions thereof to be developed, the method presented here can be extended to other countries. This is a topic for further research. However, in order to obtain a first impression of the degree of spatial differentiation to be expected for Western and Central Europe, estimates for the average values $I_{country}$ for countries other than Germany were derived. For primary pollutants, the short-range contributions were estimated according to

$$I_{country, near} = I_{D, near} \times (\rho_{country}/\rho_D) \times (u_D/u_{country}) \qquad (3.26)$$

with

$\rho_{country}$ average population density for the respective country
ρ_D average population density for Germany (230 persons/km^2, BBR 1998a)
u_D annual mean wind speed for Germany (3,5 m/s, Gerth and Christoffer 1994)

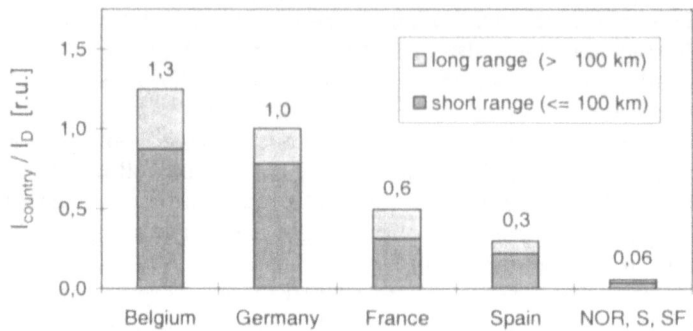

Fig. 3.18. Country average values $I_{country}$ of the population exposure per emitted mass due to traffic emissions of acetaldehyde for a selection of countries illustrating the degree of spatial variation in Western Europe. $I_{country}$ is normalized to the average value for Germany, which is $I_D = 0.62$ persons $(\mu g/m^3)$ a/kg.

$u_{country}$ annual mean wind speed for the respective country, determined as a mean value of the measurement stations for the country listed in the European Wind Atlas (Troen and Petersen 1989).

The scaling of $I_{country, near}$ with the inverse of the annual mean wind speed corresponds to the approximate proportionality of the short-range population exposure to the annual mean wind speed. Furthermore, it assumes that the shape-parameter k of the Weibull frequency distribution of short-term wind speeds and the meteorological regression formula used to the derive the combined frequency distributions of short-term wind speeds and atmospheric stability classes (see section 3.4.1.2) can be transferred to other countries. The long-range contributions for primary pollutants and the total population exposures for sulfate and nitrate aerosols were calculated by running the Windrose Trajectory Model for several emission sites spread evenly over the area of the country. Overall, the model was run for 29 emission sites (i.e. Eurogrid squares with an extension of about 100 km × 100 km) spread across Western and Central Europe (comprising the countries of the European Union and the European Free Trade Association).

Fig. 3.18 shows the country average values for traffic emissions of acetaldehyde for a selection of countries that illustrates the degree of spatial differentiation across Western and Central Europe. The country average values are likely to be similar to the low impact values for their rural areas, as it is the case for Germany. The high impact values for large cities in agglomerated regions in other countries can be expected to be similar to those for Germany, i.e. in the order of about $3 \times I_D$ in the example shown in fig. 3.18. The overall spread between the highest and the lowest class-averaged impacts can therefore be expected to be in the order of a factor $3/0.06 = 50$ for primary pollutants with short atmospheric residence times such as acetaldehyde. For Diesel particles, an analogous estimates yields a lower spread of a factor 8, which is due to their longer atmospheric residence time of about 5 days. Variations of the country average values for sulfates and nitrates are in the order of a factor 10 and 30, respectively, in part also due to variations in the background concentration of ammonia.

Table 3.4. European average fate factors F and population exposures per mass of pollutant I for primary particulates and secondary aerosols determined with the Windrose Trajectory Model (IER 1998) for 29 emission sites spread evenly across Western and Central Europe

primary pollutant	secondary pollutant	fate factor F $[m^2a/m^3]$	I [persons $(\mu g/m^3)$ a/kg]
PM 10		2,2 E-6 (long-range)	0,193 (long-range)
PM 2,5		2,0 E-5[a] (long-range)	0,746[a] (long-range)
NO$_x$	nitrate aerosol	2,9 E-6 (total)	0,191 (total)
SO$_2$	sulfate aerosol	1,5 E-6 (total)	0,109 (total)

[a] including an estimated contribution from outside of the modeling area (see section 3.4.2.3).

In the case study in chapters 4 and 5, population exposures to primary and secondary particles also need to be determined for some emissions occurring outside of the EcoSense modeling area. For this purpose, European average fate factors were determined as averages of the 29 runs of the Windrose Trajectory Model mentioned above. These fate factors can be multiplied with estimated effective population densities to yield population exposures per emitted mass according to equation (2.18). Table 3.4 also lists the corresponding average European population exposures, which will be used as default values for processes with unspecified location.

3.7
Summary and Conclusions

A method to capture the spatial differentiation of the health impacts of airborne pollutants within Life Cycle Impact Assessment was presented. The method is simple enough in terms of the required input data to be applied to the large number of emissions typically contained in a Life Cycle Inventory. It is applicable to primary air pollutants for which a linear exposure response function can be assumed and to secondary sulfate and nitrate aerosols. The consideration of the spatial differentiation of the impacts of ozone formed from nitrogen oxides and volatile organic compounds is a topic for further research. The linearity assumption represents a limitation, but nevertheless covers the most relevant primary air pollutants from energy transportation and surface transportation (particulate matter, organic carcinogens, nitrogen oxides, sulfur dioxide).

Under the prerequisite of a linear exposure response function, the impact of a given emitted mass of a particular primary pollutant depends on the emission height as well as the meteorological conditions and the population density around the emission source, but not on the time characteristics of the emission or the background concentration of the pollutant. With regards to the population density, the method presented here characterizes the continuous spectrum of possible distributions through a set of discrete settlement structure classes C within a country S. The settlement structure classes refer to the population density both at the district and at the region level and range from large cities in agglomerated regions down to low density rural districts in rural regions. The effective emission height h is kept as a continuously varying parameter, since it is relatively easy to determine. The substance parameters required as inputs are the dry deposition velocity v_d and the total atmospheric residence time τ_a. The incremental exposure of the population to a pollutant due to a given emission is calculated as an average for all sites within a settlement structure class C using a verified statistical reasoning, which is the main idea underlying the method. Due to the averaging of the population exposures within the settlement structure classes, the only meteorological input variable that needs to be considered is the annual mean wind speed u, which can be discretized into a small number of intervals.

The population exposures to secondary sulfate and nitrate aerosols, which are formed from emissions of sulfur dioxide and nitrogen oxide, respectively, are less sensitive to the emission height h and the settlement structure class C due to the time required for their formation. They are more sensitive to variations of the population density at a larger geographical scale. It is therefore sufficient to characterize the impacts of sulfate and nitrate aerosols by country average values.

In combination with suitable exposure-response slopes for the health effects of the airborne pollutants, the method allows for a calculation of the incremental human health impacts due to a given emission as a function of the input parameters (S, C, h, u, v_d, τ). A limited range of the input parameters was considered, as it was required for application within the case study on natural gas vehicles. The extension to the full parameter range is a topic for further research. In particular, settlement structure classes C were only defined for Germany. Country-average impacts were calculated for other European countries, based on their average population density and their annual mean wind speed u. The selection of substances relevant for the case study comprises particulate matter (PM 10 and PM 2,5), nitrogen oxides (NO_x), sulfur dioxide (SO_2), benzene, formaldehyde, acetaldehyde, benzo[a]pyrene and 1,3-butadiene as primary pollutants and nitrate and sulfate aerosols as secondary pollutants. For ozone formed from NO_x and volatile organic compounds (VOC), European average impacts will be used in the case study.

For primary pollutants from traffic emissions within Germany, the spatial differentiation of the impacts between the class of large cities in agglomerations and the class of rural districts in rural regions was found to range from a factor 2,2 for Diesel particles as an example for a pollutant with a long atmospheric residence time ($\tau_a = 5$ days) to a factor 5 for acetaldehyde as an example of a short-lived pollutant ($\tau_a = 9$ hours). Most of the variation is thereby due to the population density distribution around the source, while the wind speed has a smaller influence. Taking into account that the rural areas of Germany still have a much higher population density than some sparsely populated regions in Europe (e.g. the rural areas of Scandinavia), a variation of the impacts across the settlement structure classes in Europe of about a factor 8 for Diesel particles and about a factor 50 for acetaldehyde was estimated. With increasing emission height, the spatial differentiation decreases, e.g. from a factor 2,2 between large cities and rural regions in Germany for fine particulate matter (PM 2,5) emitted from vehicles at a height of a few meters to a factor 1,3 for an emission height of 200 meters. Variations of the country average values for sulfates and nitrates are in the order of a factor 10 and 30 across Europe, respectively.

4 Life Cycle Assessment of Natural Gas Vehicles: Inventory Analysis

4.1
Goal and Scope Definition

4.1.1
Introduction to the Case Study

Chapters 4 and 5 present a comparative Life Cycle Assessment (LCA) dealing with the energy conversion chains of vehicles fueled by natural gas, Diesel and petrol in order to demonstrate the practical feasibility and usefulness of the method for the site-dependent impact assessment of health effects of airborne pollutants developed in chapter 3. This LCA case study may be of interest to researchers in the fields of LCA methodology and of alternative transportation fuels. It is presented in two parts. Chapter 4 deals with the goal and scope definition and the energy and emission inventories. This provides the basis for the impact assessment and the interpretation of the results in chapter 5.

The introduction of spatial differentiation into Life Cycle Impact Assessment is particularly relevant for traffic emissions, and the method presented in chapter 3 is especially suitable to deal with mobile emission sources. This suggests to use a case study on transportation to demonstrate its applicability. Since the present work stands in a context of Technology Assessment[1], where the idea is to examine the consequences of new technologies in early phases of their development, the comparison of alternative and conventional fuels appears to be an interesting field of application.

Among the alternative fuels currently being discussed, natural gas was selected because its application is potentially becoming more widespread in many countries around the world, with environmental protection being one underlying motivation. Whether or not natural gas vehicles will achieve a breakthrough in countries where they have not been used on a significant scale so far (e.g. in Germany) may be decided within the next few years. At present, increased efforts in this direction are made by both natural gas suppliers and some vehicle manufacturers. Various demonstration projects on natural gas vehicles (NGVs) have been conducted, such that data on the emissions and fuel consumptions of those vehicles are available, even though their quality still needs to be improved. On the other hand, since natural gas vehicles are still few in number at present (about 3600 in

---

[1] See section 2.1 for a discussion of the relation between Life Cycle Assessment and Technology Assessment.

Germany at present, IANGV 1999), the case study still has the character of a prospective Technology Assessment. It is partial in the sense of being limited to environmental issues.

In Germany as well as in other countries, a sufficient infrastructure of fuelling stations for natural gas is still lacking. At present, there are about 80 stations in Germany (IANGV 1999), while about 300 would be required for a halfway acceptable coverage of the whole country[2]. For this reason, the main area of use of natural gas as a fuel in Germany is currently in fleets of vehicles that operate within a limited radius. In particular, demonstration programs on buses fueled by natural gas have been implemented in a number of German and other European cities out of concern about the health impacts of emissions from Diesel buses, especially of particulate matter and of nitrogen oxides (NO_x). The comparison of environmental effects of using natural gas and Diesel as fuels for city buses was therefore selected as one part of the case study. In addition to buses, the use of natural gas as a fuel for cars (e.g. taxis or small delivery vehicles) is also considered in the case study, in which case the reference fuels for comparison are Diesel and petrol.

Various instruments of Technology Assessment that relate to environmental issues, namely LCA, Risk Assessment and Environmental Impact Assessment (see section 2.1) are in principle available to address this case study. In choosing the instrument which is best suitable for this purpose, alone or in combination with others, it should be borne in mind that, while the issue of the health effects of the vehicle emissions is an important one, the use of natural gas also raises questions with regards to climate change. On the one hand, the specific CO_2 emissions of natural gas are smaller than those of Diesel and petrol (55 g CO_2/MJ vs. 74 and 75 g CO_2/MJ, Rausch et al. 1998). On the other hand, it turns out that the fuel consumption of natural gas vehicles is higher than that of vehicles fueled by Diesel or petrol, according to the higher values reported in the literature to such an extent that the advantage of smaller specific CO_2 emissions may be overcompensated. Furthermore, the frequently discussed issue of high methane losses during the extraction and pipeline transport of natural gas to the point of combustion, in particular in Russia, also needs to be considered when assessing the benefits or otherwise of natural gas as a transportation fuel in terms of climate change.

Comparing the use of natural gas, Diesel and petrol as vehicle fuels in terms of their respective environmental effects therefore requires looking at both upstream processes and vehicle emissions and considering a broad spectrum of pollutants in both cases in order to avoid possible trade-offs being overlooked. LCA with its encompassing approach (see chapter 2) therefore appears to be the appropriate instrument to address the topic of the case study.

While the benefits in terms of human health of substituting Diesel or petrol by natural gas for the propulsion of street vehicles are qualitatively obvious due to significant reductions of the emissions of particulate matter, nitrogen oxides and other pollutants, their quantification is nevertheless worthwhile for two reasons. For one, the modeling of the cause-and-effect chains from the emissions up to health-relevant endpoints allows for a comparison of the relative significance of

---

[2] For comparison, there were about 17,000 conventional fuelling stations in Germany in 1998 (Shell 1998a: 26).

the emission reductions of the various pollutants. Furthermore, it is interesting to examine the dependence of the health benefits on the location where the fuel substitution takes place, which goes beyond the generic, i.e. site-independent considerations typical for LCA. This can be helpful in terms of prioritization of measures of emission reduction, i.e. in allocating sparse resources for environmental protection in the most efficient way. In this regard, a site-dependent quantification of the health impacts allows to examine the frequently made implicit or explicit assumption that the use of natural gas is only worthwhile in large cities.

This assumption appears to be plausible at first sight. It may be motivated by the fact that possible exceedances of legal concentration limits, e.g. of particles, nitrogen oxides and benzene according to the 23rd Amendment to the German Air Pollution Control Act (23. Bundes-Immissions-Schutz-Verordnung, see, e.g., Seidler 1994), are indeed only of concern within urban street canyons. However, health effects of these air pollutants also occur at concentrations below the legal limits (section 2.5.2.5) and therefore cannot be captured by considering the pollutant concentrations at a few individual ‚hotspots‘ only.

Considerations of distributional justice may be another possible motivation for the above assumption. Accordingly, emission reductions within cities might be given higher priority due to the higher background concentrations. However, a quantification of that idea would involve contentious normative presuppositions and would furthermore require a detailed allocation of the background concentrations to specific parts of the population. Furthermore, the distribution of other environmental pressures would then have to be considered as well. Instead, the total health effects resulting from an integration over the relevant area (equation 2.12), on which the method presented in chapter 3 is based, appear to be a more reasonable measure.

In examining the influence of the location of use of the vehicles in the above sense, it is not so much *specific* locations that are of interest. Rather, in a first round of investigation, it is more suitable to consider *typical* situations, e.g. the typical large city or the typical small municipality in Germany. This is because it is impossible to investigate all cases (several thousand municipalities in Germany) in detail, while the selection of particular examples always bears the risk of not being representative. The method of defining generic spatial classes of emission sites presented in chapter 3 offers a spectrum of such typical situations, the representativity of which has been verified, and is therefore well suited to examine the influence that the location of use of the vehicles may have on the outcome of the comparison of the fuels in a quantitative way.

4.1.2
System Boundaries

The encompassing approach of LCA implies that a comparison of different types of vehicles should, besides the emissions from the vehicles themselves, also consider emissions associated with the supply chain of the fuels (‚fuel chain‘), the manufacturing, maintenance and disposal of the vehicles as well as the construction of the infrastructure. The relative contribution of each of these four aspects to the total emissions of airborne pollutants (per kilometer of driving) is shown in

Table 4.1. Relative contribution [%] of (1) vehicle operation, (2) upstream fuel supply chain, (3) construction, maintenance and disposal of the vehicles, and (4) construction of the infrastructure to the emissions of selected airborne pollutants through the operation of city buses (EURO 2 emission control standard) in Switzerland (Maibach et al. 1995: 271)

pollutant	(1) operation	(2) fuel supply chain	(3) manu- facture	(4) infra- structure	total
CO_2	74	10	7	8	100
CH_4	2	68	13	16	100
N_2O	70	9	8	13	100
NMVOC	26	55	6	13	100
SO_x as SO_2	26	33	23	18	100
NO_x as NO_2	84	8	2	6	100
particles	12	21	32	34	100

table 4.1 for the case of city buses using Diesel fuel in Switzerland (Maibach et al. 1995: 271).

Accordingly, the use phase of the vehicles dominates the emissions of most air pollutants, while the emissions of methane (CH_4) and non-methane volatile organic compounds (NMVOC) are dominated by the upstream fuel chain[3]. For oxidized sulfur compounds (SO_x) and particles, the contributions of the four phases are of similar magnitude. It should be noted, though, that this refers to a rather unspecified pollutant category of particles. Besides combustion products, it also contains coarse particles such as dust from raw material extraction and processing. The coarse particles, which are not relevant to human health, constitute a significant share (90 % and more) in the case of the construction of the infrastructure and the vehicles (Friedrich et al. 1998:55). The relative contribution of the vehicle operation to the emissions of the combustion products relevant to human health (here in particular Diesel soot particles) is therefore actually much higher.

With regards to the present comparative LCA of vehicles fueled by natural gas, petrol and Diesel, the infrastructure does not need to be considered since its contribution, in addition to being small for most air pollutants, is the same in all cases and therefore cancels out when the difference between the fuels is considered.

The manufacturing of the vehicles is also very similar in all cases. Natural gas vehicles typically are modifications of standard Diesel or petrol vehicles. The modifications concern some parts of the engine and the fuel system. The most significant modification can be expected to be associated with the pressurized gas tanks. Their additional mass depends on the range of the vehicles and the materials they are made of. For buses with a range of about 280 kilometers and composite steel tanks wrapped with aramid fibres, the additional mass amounts to about 13 % of the total mass of the buses (Bartosch 1997:67-68; Rösgen et al. 1997:9-12). A similar relative additional mass applies to natural gas cars. Under the assumption of similar emissions per mass of output for the manufacturing of the gas tanks than for the rest of the bus, which is plausible because of the large fractions of steel in

---

[3] According to the results presented below (tables 4.19 and 4.21), the NMVOC emissions are also dominated by the use phase.

both cases, the contribution of the gas tanks to the total emissions can be expected to be below 4 % according column (3) of table 4.1. The differences in the manufacturing stages of the vehicles will therefore be neglected here. Like the infrastructure, the manufacturing will therefore not be considered in the following, since it approximately cancels out in the difference between the options. However, the effect of the additional mass of natural gas vehicles on their fuel consumptions and use phase emissions is taken into account.

This leaves the use of the vehicles and the respective fuel supply chains to be considered. Together, these will be referred to as *energy conversion chains*. The fuel supply chains will also be referred to as *upstream processes*. While emissions of the upstream processes are smaller than the vehicle emissions for most of the air pollutants shown in table 4.1, they can nevertheless not be neglected a priori since they differ between Diesel, petrol and natural gas.

4.1.3
Selection of Pollutants and Impact Categories

As mentioned above, climate change and human health are two issues of particular interest when comparing natural gas, Diesel and petrol vehicles. The selection of further impact categories to be considered was guided by the fact that the main emission reductions from the substitution of Diesel or petrol by natural gas are expected for particulate matter (PM), non-methane volatile organic compounds (NMVOC), including a number of carcinogenic substances, nitrogen oxide (NO_x) and sulfur dioxide (SO_2). The latter two pollutants are important acidifying agents. NO_x is also involved in nutrification and, in combination with NMVOC, in photo-oxidant formation. Furthermore, the extraction of abiotic resources is an important category when dealing with the use of fossil fuels. Taken together, the following of the impact categories from the standard list in section 2.5.1 will be considered in the case study:

1. Extraction of abiotic resources (here: deposits)
2. Climate change
3. Human toxicity
4. Photo-oxidant formation
5. Acidification
6. Nutrification.

Among the impact categories not considered here are land use and eco-toxicity. Impacts of this kind can be expected mainly from the extraction of raw oil and natural gas. In the case of off-shore extraction, in particular, chemicals are released to the sea water which may have an effect on the aquatic biota (Karman and Reerink 1997). Furthermore, the extraction may also conflict with the preservation of natural habitats (Marquenie and Verburgh 1997). However, the operations of oil and gas production are very similar and can therefore be generally be expected to lead to similar impacts. In fact, sometimes both crude oil and natural gas are extracted at the same site. For this reason, neglecting the impact categories of land use and eco-toxicity appears to be a good approximation when considering the differences between the three fuels. The third impact category not considered here is stratospheric ozone depletion, since the energy conversion chains of natural gas

Table 4.2. Air pollutants and their characterization in terms of impact categories, as considered in the case study

pollutant		impact category				
		climate change	human toxicity	photo-oxidants	acidifi-cation	nutrifi-cation
1	carbon dioxide (CO_2)	X				
2	methane (CH_4)	X		X		
3	nitrous oxide (N_2O)	X				
4	particulate matter (PM)[a]		X			
5	nitrogen oxide (NO_x)		X[b]	X	X	X
6	sulfur dioxide (SO_2)		X[b]		X	
7	NMVOC			X[c]		
8	benzene (C_6H_6)		X[d]			
9	formaldehyde (CH_2O)		X[d]			
10	acetaldehyde (C_2H_4O)		X[d]			
11	benzo[a]pyrene (B[a]P)		X[d]			
12	1,3-butadiene (C_4H_6)		X[d]			

[a] differentiated into particles with diameters < 2,5 μm (PM 2,5) and < 10 μm (PM 10).

[b] Only secondary nitrate and sulfate aerosols are considered (see section 2.5.2.5).

[c] The sum-parameter NMVOC is used for the assessment of tropospheric ozone formation, while impacts on human health are considered for the individual carcinogenic NMVOC (positions 8-12).

[d] The upstream emissions of these carcinogenic pollutants are neglected (see text for justification).

and Diesel are not associated with significant emissions of the respective halogenated organic compounds (Rausch et al. 1998).

Since the main environmental impacts of energy conversion chains are associated with airborne pollutants from combustion processes, the analysis will be limited to airborne pollutants. The pollutants considered here and their association with the five output related impact categories of interest here are shown in table 4.2. The selection of the organic compounds (positions 8-12) is based on (Mangelsdorf et al. 1999) and refers to their contribution to the carcinogenic effects of the vehicle exhausts. Their upstream emissions can be neglected. This can be justified on the grounds that their vehicle emissions only make a small contribution to the overall health effects (sections 5.1.1.3 and 5.1.1.4), and that their upstream emissions are likely to be smaller, given that the upstream NMVOC emissions are generally smaller than the NMVOC emissions of the vehicles (sections 4.3.2 and 4.4.2).

With regards to the formation of photo-oxidants, other NMVOC contained in the exhaust gases would generally have to be considered with higher priority. However, due to a lack of data on the respective emissions factors, the contribution of individual substances within the group of NMVOC to photo-oxidant formation is not considered here. Photo-oxidant formation is therefore assessed on the basis of the aggregate NMVOC emissions.

4.1.4
Functional Unit

The functional unit considered here is one vehicle-kilometer of inner-city driving with a city bus or a car in Germany[4]. Since the interior space of a natural gas bus is not affected by the additional pressurized tanks, which are typically mounted on the roof, the average occupation of natural gas buses and Diesel buses is assumed to be identical. A separate comparison on the basis of one person-kilometer is therefore not necessary. The same essentially applies to cars as well, with the exception that the storage space for luggage may be somewhat affected. For the uses of cars considered here, e.g. as taxis, luggage space is generally not a limiting factor, such that the natural gas cars may also be considered as functionally equivalent to their petrol and Diesel counterparts.

As was discussed above, a specification of the type of municipality in which the vehicles are driving will be added to the definition of the functional unit. This differentiation is made with regards to the dependence of the health impacts associated with a given emission on the population density in the area around the source (see chapter 3). In this regard, three different cases are distinguished, with the shorthand notations for the respective settlement structure classes indicated in brackets as defined in section 3.3.1:

1. large city in agglomerated region (I,1) (briefly denoted as „large city")
2. central city in urbanized region (II,1) („normal city")
3. small city in urbanized region (II,3) („small city").

The driving patterns and the associated emissions of the vehicles are assumed to be the same in all three cases. Hence, the differentiation between the three cases will only be used for the impact assessment in chapter 5.

The emission inventory will be presented in two parts. In section 4.2, the upstream processes will be considered which are associated with providing a given amount of fuel (1 TeraJoule = 10^{12} Joule) at a fuel station. These data will be combined with the fuel consumptions and emissions of the buses and cars in sections 4.3 and 4.4, respectively, to yield the overall emissions associated with one kilometer of driving.

The inventory data presented here were, for the most part, calculated with the GEMIS software, version 3.08 (Rausch et al. 1998). GEMIS combines an encompassing database on processes of energy conversion with an algorithm to calculate emission inventories for energy conversion chains. It is one of two commonly used standard references for LCI data on energy technologies, the other one being (Frischknecht et al. 1996). The GEMIS software tool was used here because it allows the user to define new processes. This makes it possible to alter some of the existing data and to incorporate new datasets. For the present case study, such alterations were made in particular with regards to the following points:

---

[4] With regards to the calendar time, the one kilometer of driving is supposed to represent an average over the entire lifetime of the vehicles, since this is the period of interest when considering an investment in a new technology such as natural gas vehicles. Therefore, the use of annual mean rather than short-term pollutant concentrations in the impact assessment method presented in chapter 3 is appropriate for an application in the case study (chapter 5).

- More recent data for the year 1998 were used for the import shares of natural gas and raw oil from various countries to Germany.
- Two options were considered for the electricity demand of the compression of the natural gas at the fuel stations in order to represent the spectrum between small and large fuel stations.
- Emissions of non-methane volatile organic compounds (NMVOC) due to losses of natural gas were considered in addition to those of methane.

4.2
Fuel Supply Chains

The fuel supply chains comprise all processes from the exploration and extraction of the resources to the distribution of the fuel to a fuel station and include any intermediate steps of processing and transportation. All inventory data provided in this section refer to an amount of 1 TeraJoule (1 TJ = 10^{12} Joule) of fuel provided at a fuel station in Germany. On this amount of fuel, an average Diesel city bus can drive about 83,000 kilometers, which is in the order of the distance typically driven within one year.

4.2.1
Natural Gas

4.2.1.1
Input Data

The natural gas consumed in Germany is, for the most part, imported from other countries, with around one fifth coming from domestic extraction. Based on the market shares for the year 1998[5], the natural gas supply in Germany was modeled as a mix of 35 % imports from Russia, 22 % each imports from the Netherlands and Norway (including 3 % from Denmark and the United Kingdom) and 21 % extraction in Germany (Ruhrgas AG 1999). The percentages thereby refer to the lower calorific value H_u of the fuel.

For each of these countries of origin, the extraction, processing and transport of the natural gas is essentially modeled in the same way within the GEMIS database (Rausch et al. 1998; see fig. 4.1): The energy used during the exploration phase can be neglected. For the extraction of the gas, electrical motors are used which are supplied with electricity either by a gas power station located near the site of extraction (Russia, Norway) or by the national grid (Netherlands, Germany). Depending on its composition, which varies from country to country, the extracted gas is dried and desulfurized, and hydrocarbons with higher molecular weights are separated off. These processing steps require heat and electricity, which are provided by a gas burner and the national power grid, respectively. The natural gas from Russia, Norway and the Netherlands is subsequently transported to Germany

---

[5] These shares are somewhat different from the GEMIS default data (Rausch et al. 1998), which represent a forecast based on an analysis of the long-term supply contracts of the natural gas industry.

through pipelines. The compressor stations providing the necessary transport energy are powered by natural gas.

Within Germany, the natural gas is further distributed regionally and locally through pipelines at different pressure levels. While regional transmission typically occurs at gas pressures of 15-40 bar, the pressures in the local distributions grids range from about 10 bar at city gate stations down to about 25 mbar for delivery to residential and small scale users (Peebles 1992:127-129).

While natural gas fuel stations can, in principle, be connected to the low pressure distribution grid for small scale users, this should be avoided for two reasons. First, the low pressure local distribution leads to relatively high losses of natural gas and corresponding methane emissions. For a pipeline of 10 kilometers in length, these losses amount to 0,7 mass-% (Rausch et al. 1998). Second, the natural gas needs to be compressed to about 200 bar at the fuel stations in order to fill the pressurized tanks of the vehicles. The energy required for the compression of the gas is significant (see tables 4.3 and 4.4) and increases with decreasing input pressure. Therefore, it is assumed in the following that natural gas fuel stations are not connected to the low pressure local distribution grid. With regards to the input pressure, two cases are distinguished:

1. A large fuel station with an input pressure of 12 bar (reference case):
   According to measurements for the natural gas fuel station in Hannover, Germany, this leads to an average electricity demand for the compression of 0,16 kWh/Nm³ (Schüle 1997a). This is equivalent to 0,017 TJ electricity / TJ natural gas, using the lower calorific value of 33,8 MJ/Nm³ of natural gas in Germany (Rausch et al. 1998). This electricity demand is at the lower end of the range of values for fuel stations in Germany (Schüle 1997b:fig. 3).

2. A small fuel station with an input pressure of 1,3 bar (for sensitivity analyses):
   Compared to the large fuel station, this input pressure leads to a twofold electricity demand for the compression (0,034 TJ electricity / TJ natural gas), which is at the higher end of the range of values for natural gas fuel stations in Germany (Schüle 1997b: fig. 3).

The amounts of energy required for each upstream process are shown in table 4.3 together with the respective losses of natural gas, which represent emissions of its constituents methane (CH_4) and non-methane volatile organic compounds (NMVOC)[6]. As can be seen in table 4.3, differences between the four countries of extraction mainly result from the different lengths of the pipelines. The long transport distance (7000 kilometers) for the gas imported from Russia leads to a high gas demand for the compressors. In combination with higher loss rates of natural gas per kilometer of pipeline, it also leads to significantly higher methane

---

[6] In the GEMIS database (Rausch et al. 1998), the NMVOC emissions due to the losses of natural gas are neglected. However, these NMVOC emissions are larger than those occurring at the power plants providing the auxiliary energies. The GEMIS default data were therefore altered to include the NMVOC losses, which were calculated by combining the loss ratios listed in table 4.3 with the regionally varying chemical composition of the natural gas. Emissions of possible other constituents of natural gas such as nitrogen, hydrogen, carbon dioxide, carbon monoxide and hydrogen sulfide were neglected either because of their small share of the overall mass or because they are not harmful to human health or the environment.

Table 4.3. Relative energy inputs and losses of natural gas for the supply of compressed natural gas at a fuel station in Germany (Rausch et al. 1998)

		natural gas from			
		RUS	NOR	NL	D
market share in Germany [%][a]		35	22[b]	22	21
extraction					
electricity	[% of H_u]	0,11[c]	0,11	0,11	0,11
losses of natural gas	[%]	0,5	0,2	0,125	0,075
processing					
electricity	[% of H_u]	0,2	0,1	0,1	0,5
process heat	[% of H_u]	0,64	0,1	0,1	0,5
losses of natural gas	[%]	0,25[d]	0,125	0,125	0,075
pipeline transport to Germany					
length of pipeline	[km]	7000	1700	600	/
gas for compressors	[% of H_u]	4,12	0,87	0,37	/
specific losses of natural gas	[%/100km]	0,016	0,0006	0,0006	/
absolute losses of natural gas	[%]	1,12	0,0102	0,0036	/
regional distribution in Germany					
length of pipeline	[km]		250		
gas for compressors	[% of H_u]		0,1347		
specific losses of natural gas	[%/100km]		0,0006		
absolute losses of natural gas	[%]		0,0015		
compression at fuel station:					
electricity, large station	[% of H_u]		1,7041		
(electricity, small station)	[% of H_u]		(3,4082)		

H_u lower calorific value
[a] in 1998 (Ruhrgas AG 1999).
[b] including 3 % for Denmark and the United Kingdom.
[c] The probably erroneous value of 0,1 % in the GEMIS database (Rausch et al. 1998) was corrected to be consistent with the other countries.
[d] The unexplained value of 95 % for the efficiency of the processing step (Rausch et al. 1998) was altered to 100 % to be consistent with the other countries.

emissions. The overall loss rate of 1,87 mass-% for natural gas imported from Russia (Rausch et al. 1998) used here is in accordance with an estimate of 1,82 mass-% by Zittel (1997). Based on an extrapolation of measurements at selected facilities to the entire gas production in Russia, a more recent study yields a comparable loss rate of (1 ± 0,5) mass-% (Dedikov et al. 1999). In accordance with the values used here (table 4.3), most of the losses (0,9 mass-%) are thereby attributed to pipeline transport. The loss rate for extraction and processing was found to be 0,1 mass-%, which is lower than the value of 0,75 mass-% used here.

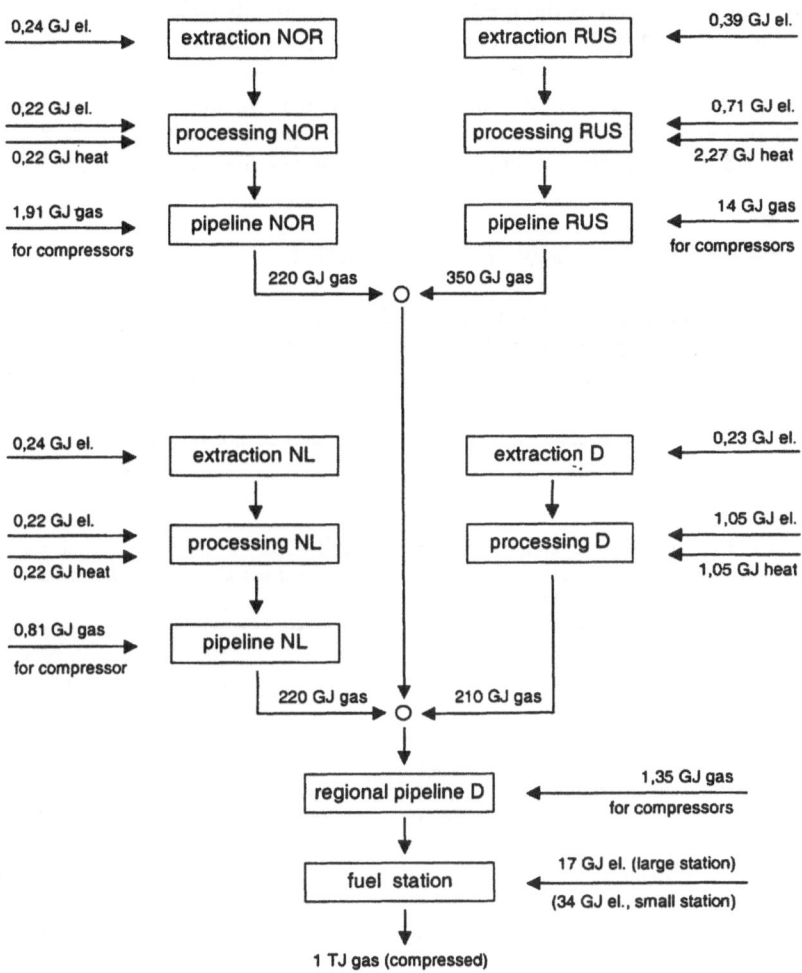

Fig. 4.1. Main energy flows associated with the supply of compressed natural gas at a fuel station in Germany according to the data in table 4.3. *el.* electricity

Leaks and intentional releases required for maintenance or repairs were identified to be the main causes for the losses.

The structure of the process chain associated with the import of natural gas from the four countries of origin is shown in fig. 4.1. In this figure, the absolute values of the energy flows, which result from combining the relative values given in table 4.3 with a supply of 1 TeraJoule of natural gas to a fuel station, are also indicated. It should be noted that fig. 4.1 only shows a top layer of energy flows. Each of the processes delivering energy to the exploration, processing and transport of the natural gas is associated with further upstream processes which are not shown. Among these are the provision of the fuels to the power stations, compressors or burners as well as the production of the materials from which these facilities as well as the pipelines are constructed. While not shown in the graphs, these processes are taken into account in the calculations. However, since it is only in a few cases that they contribute significantly (with more than 1 %) to the overall energy demands and emissions, their associated input data, which can be found in the GEMIS database (Rausch et al. 1998), are not described here in detail.

4.2.1.2
Emissions and Cumulative Energy Demand

Based on the input data given in table 4.3, the cumulated emissions and the Cumulative Energy Demand (CED) associated with the provision of 1 TJ of natural gas at a natural gas fuel station in Germany were calculated using the GEMIS software. The calculations proceed such that the processes upstream of those shown in fig. 4.1 are traced backwards until processes are reached that do not put

Table 4.4. Emissions and Cumulative Energy Demand (*CED*) associated with the supply of 1 TeraJoule (lower calorific value H_u) of compressed natural gas at a fuel station in Germany - **Part 1:** Pollutants with regional range of dispersion

	PM [g]	SO$_2$ [g]	NO$_x$ [g]	NMVOC [g]
gas from NL (H_u=220,0 GJ)				
extraction	44,1	56,7	160,2	428,8
processing	63,3	81,1	233,3	432,4
pipeline transport	11,1	11,9	266,4	23,2
gas from D (H_u=210,0 GJ)				
extraction	13,2	40,0	86,2	73,9
processing	47,9	163,5	437,7	108,8
gas from NOR (H_u=220,0 GJ)				
extraction	20,5	27,3	244,4	1906,8
processing	32,6	41,9	115,0	1188,3
pipeline transport	28,7	32,3	606,3	141,4
gas from RUS (H_u=350,0 GJ)				
extraction	31,0	27,6	572,4	629,1
processing	381,5	1187,6	917,5	341,5
pipeline transport	308,4	316,8	5242,3	1666,8
distribution in D (H_u=1,0 TJ)				
pipeline transport	20,4	23,8	434,5	28,4
compression at large fuel station	270,0	2016,0	4380,5	437,5
(compression at small fuel station)	(540,0)	(4032,0)	(8761,1)	(875,0)
SUM for large fuel station	1272,8	4026,4	13696,6	7406,9
(SUM for small fuel station)	(1542,8)	(6042,4)	(18077,1)	(7844,4)

Table 4.4. Emissions and Cumulative Energy Demand (CED) associated with the supply of 1 TeraJoule (lower calorific value H_u) of compressed natural gas at a fuel station in Germany - **Part 2:** Pollutants with hemispherical or global range of dispersion and CED

	CO_2 [kg]	CH_4 [g]	N_2O [g]	CED-H_u [GJ]	CED/H_u-1 [%]
gas from NL (H$_u$=220,0 GJ)					
extraction	48,0	5136,9	1,9	0,7	0,3
processing	90,4	5241,4	2,6	1,2	0,5
pipeline transport	53,5	231,1	2,2	0,9	0,4
gas from D (H$_u$=210,0 GJ)					
extraction	53,3	3291,0	1,8	0,8	0,4
processing	297,5	3919,6	9,1	4,7	2,2
gas from NOR (H$_u$=220,0 GJ)					
extraction	67,8	7171,7	2,3	1,0	0,5
processing	59,0	4554,8	1,1	0,9	0,4
pipeline transport	127,1	558,2	5,0	2,2	1,0
gas from RUS (H$_u$=350,0 GJ)					
extraction	108,3	34655,8	4,3	1,9	0,5
processing	406,8	18274,1	10,8	6,4	1,8
pipeline transport	957,1	79469,0	38,4	21,4	6,1
distribution in D (H$_u$=1,0 TJ)					
pipeline transport	88,0	579,1	3,6	1,6	0,2
compression at large fuel station	3070,0	7314,6	116,5	50,1	5,0
(compression at small fuel station)	(6140,0)	(14629,3)	(232,9)	(100,1)	(10,0)
SUM for large fuel station	5426,8	170397,3	199,6	93,8	9,4
(SUM for small fuel station)	(8496,8)	(177712,0)	(316,1)	(143,9)	(14,4)

further demands on other processes. Alternatively, loops may occur. For example, the power stations delivering electricity to the extraction of natural gas are themselves fueled by natural gas. An iterative calculation procedure is implemented in GEMIS to deal with such loops. The iteration stops once the relative difference between the results of two steps is smaller than a preselected ε (Fritsche et al. 1994:224). The calculations were carried out with an accuracy of $\varepsilon = 0,01$ %.

The Cumulative Energy Demand (CED) and the cumulated emissions of CO_2, CH_4, N_2O, PM, SO_2, NO_x and NMVOC associated with the supply of 1 TJ of natural gas at a fuel station in Germany are shown in the last row of table 4.4. The table also contains the cumulated emissions and energy demands associated with each phase (extraction, processing, transport, compression) of the fuel supply chain. These contributions to the overall sum were calculated by defining separate processes within the GEMIS software[7]. Since the Cumulative Energy Demand (CED) also contains the lower calorific value H_u of the supplied fuel itself, table 4.4 lists the differences CED-H_u (in GJ as well as in % relative to H_u) as a more reasonable measure of the energy demand associated with the supply of the fuel which does not include the burning of the fuel.

The absolute values of this difference, as shown in the fourth column of table 4.4, part 2, can serve as a proxy indicator for the environmental significance of the various phases of the fuel supply chain. The highest value is attributed to the compression of the gas at the large (small) fuel station amounts to 5 (10) % of the energy content of the supplied fuel. In comparison, extraction requires 0,4 %, processing 1,3 % and pipeline transport 2,4 % of the energy content of the supplied fuel for all four countries taken together. The environmental relevance of the compression is also apparent in the fact that, even for the more favorable large fuel station, the associated emissions assume the first or second and in one case third rank among the phases listed in table 4.4 for all pollutants except methane and non-methane volatile organic compounds (NMVOC)[8].

Table 4.4 furthermore shows that the import of natural gas from Russia is associated with a high CED (6,1 % of H_u) and high emissions per unit of gas. This is mainly due to the long transport distance of 7000 kilometers. In the case of the methane emissions, this is aggravated by the higher specific loss rate of natural gas.

4.2.1.3
Spatial Disaggregation

Input data: In order to combine the emission inventory contained in table 4.4 with the site-dependent impact indicators derived in chapter 3 for the pollutants PM, SO_2 and NO_x, the inventory for these pollutants was further disaggregated according to the different locations and heights of the contributing processes. As was discussed in section 2.5.2.5, SO_2 and NO_x will only be considered in terms of their associated secondary sulfate and nitrate aerosols. Within the approximations of the Windrose Trajectory Model (IER 1998) used here, the concentrations of these secondary pollutants are independent of the emission height. The emission height therefore only needs to be considered for primary particles. In terms of their size fractions, the emitted masses contained in the GEMIS database were interpreted as particles with a diameter of less than 10 μm (PM 10) due to their anthropogenic origin (Bliefert 1995:215) for all processes except emissions from stationary or

---

[7] The values in the last row of table 4.4 are the sums of the emissions and energy demands for the individual phases. These sums may differ by a few 0,01% from the emissions and energy demands calculated by GEMIS for the entire fuel supply chain. These numerical deviations corresponds to the relative convergence criterion $\varepsilon = 0,01$ % in the case of loops in the product system. They are due to differences in the number of iterations required to reach the convergence relative to the different emissions or energy demands for the entire fuel supply chain and parts thereof, respectively.

[8] The situation for the methane and NMVOC emissions is different because they result, for the most part, not from combustion processes but from direct losses of natural gas.

mobile Diesel engines, which were characterized as PM 2,5, i.e. particles with diameters of less than 2,5 μm[9].

Since the determination of the emission height and the settlement structure type or population density around the emission source requires some time and effort, it cannot be carried out for all the processes covered by the inventory, of which there are several hundred in this case. Therefore, the spatial disaggregation was only carried out for those processes that contribute more than 1 % to the overall cumulated emission for a particular pollutant.

The choice of this cutoff criterion can be justified on the grounds that processes contributing less than 1 % to the overall emissions are unlikely to contribute significantly to the overall impacts, such that a site-dependent assessment of their impacts is not going to lead to new insights. Furthermore, since there is typically a large number of such "small" processes spread across a more or less random sample of locations, spatial variations of their impacts per emitted mass are likely to cancel each other out to a large extent. Hence, the impact of all the "small" processes taken together can, in good approximation, determined by using average values for emission height and population density. Since most of the processes are located in Europe, European average population exposures per mass of pollutant were used for this purpose (table 3.4).

Tables 4.5, 4.6 and 4.7 show the disaggregated emission inventories for the pollutants PM, SO_2 and NO_x. In each of these tables, the first column contains the emissions associated with the various life cycle stages as given in table 4.4, part 1. The columns 2 to 7 contain information about those individual processes within these stages that contribute more than 1 % to the sum over the whole life cycle. As it turns out, these values add up to between 73 and 84 % of the total emitted mass. Hence, the cutoff criterion is reasonable in that it captures the largest share of the overall emissions while the number of individual processes to be considered still remains reasonable (between 8 and 14 in this case).

For the most part, the processes contributing more than 1 % to the overall emissions are those that supply the energy for the extraction, processing, transport, distribution and compression of the natural gas according to the energy demands listed in table 4.3. In some cases, processes upstream of the respective power plants, compressors or burners also contribute significantly. This is the case for the PM emissions from the power plant mix in Russia providing electricity for (a) the processing of the natural gas that powers the compressor stations for the pipeline transport of natural gas from Russia to Germany and (b) the processing of natural gas and the transport of raw oil from Russia that power (in part) the power plant mix in Germany providing electricity to the compression of the natural gas at the fuel stations (table 4.5). It is also the case for the SO_2 emissions from the oil burners in Russia that are used for the processing of raw oil that is used as fuel in power plants in Russia for the processing of the natural gas and in power plants in Germany for the compression of the gas (table 4.7).

---

[9] The characterization of particles as either PM 10 or PM 2,5 refers to the associated epidemiological effect factors (table 2.7). For the purpose of dispersion modeling with the Windrose Trajectory Model, which uses a different size classification, PM 2,5 were treated as particles with diameters < 0,95 μm, and PM 10 as particles in the diameter range 0,95-4 μm in order to obtain reasonable values of the atmospheric residence times (see table A.1 in the appendix).

In some other cases, the production of the materials (steel, cement) used as construction materials for certain facilities contributes more than 1 % to the overall emissions. For the PM emissions, this applies to the plants for the processing of gas in Russia and the gas pipeline from Russia to Germany (table 4.5). For the SO_2 emissions, it applies to the pipeline from Russia to Germany and to the ship transport of iron ore from which the steel for this pipeline and for the power plant mix in Germany providing electricity for the compression is made, i.e. from a process upstream of the materials production itself (table 4.7).

Tables 4.5 to 4.7 contain information on the location and the effective emission height of the significant individual processes. These data are not usually contained in Life Cycle Inventories, i.e. they represent additional data required to introduce spatial differentiation into the Life Cycle Impact Assessment of health effects of airborne pollutants according to the method presented in chapter 3[10].

With regards to acquiring information about the location of individual emissions in the inventory, the GEMIS database is useful in that the country in which the emission takes place is given for each unit process. In the case of emissions from high stacks (h $\approx$ 200 m, such as from power plants or large chemical facilities), this information, together with the average population density of the country, is sufficient to determine the population exposure since no significant differentiation between rural areas and large cities occurs (fig. 3.14).

In the case of lower emission heights, further information on the location of the source needs to be added. In the case of traffic emissions, the standard inventory data can sometimes be helpful in this regard: traffic emissions are typically differentiated according to whether they occur in cities, on rural roads or on highways since the different driving conditions influence the mass of the emissions. This differentiation can, at least approximately, be expressed in terms of the settlement structure classes introduced in chapter 3 and thereby be employed for the purpose of a spatially differentiated impact assessment as well.

The effective emission heights listed in tables 4.5 to 4.7 represent rough estimates. All power plant emissions were assumed to occur through high stacks (in the order of 200 meters), while traffic emissions (from ships in this case) can be assumed to have a low emission height in the order of 5 meters. For the remaining processes (compressor stations along pipelines, materials production), a medium emission height in the order of 50 meters was assumed. In most cases, such rough estimates are sufficient since they are combined with an essentially constant radial population density distribution, in which case the influence of the emission height is small (fig. 3.14).

For emission sites within Europe, the population exposures per mass of emitted pollutant listed in tables 4.5 to 4.7 were calculated as described in chapter 3: The Ecosense software was used to calculate the total population exposure to secondary sulfate and nitrate aerosols and the long-range contribution to the population exposure for primary particles (section 3.4.2). The short-range contribution to the latter was taken to be equal to the default value for the country of emission and the

---

[10] Data concerning the location of processes may sometimes be available, for example if they are required to determine transport distances, but emission heights are presently not considered in Life Cycle Inventories.

appropriate emission height (see section 3.6.6), unless indications were available that the emission site is located in more densely populated areas such as cities.

For emission sites outside of Europe (here: Russia), European average fate factors (table 3.4) were combined with estimated effective population densities (see below) to obtain the population exposures per mass of pollutant. In the case of primary particles, this was done for the long-range only, with the short range contributions calculated in the same way as for emission sites within Europe. For the sum of the contributions < 1 % of the total emitted mass, the European average population exposures per mass of pollutant given in table 3.4 were used. In the case of primary particles, this was done for the long-range only, while the short-range contribution was calculated for the average population density of the land masses in Western and Central Europe of 115 persons/km^2. Additional, more detailed comments on the assumed locations and emission heights of individual processes in tables 4.5 to 4.7 are provided in the following.

The steel used for the construction of the natural gas processing plant in Russia and the respective pipelines was assumed to be supplied from domestic production. Therefore, an effective population density equal to the average population density of 9 persons/km^2 (Hofstetter 1998:243) was used for these as well as all other processes occurring in Russia. The offshore pipeline transport of natural gas from Norway to Germany was assumed to be effected by one compressor station located at Kollsnes in Norway, for which the population exposures (both in the short and in the long range) were calculated with the EcoSense software. The onshore pipeline transport from the Netherlands to Germany was treated in the same way. For the long pipeline transport of natural gas from Russia (West Siberia) to Germany, the effective population density for the numerous compressor stations was assumed to vary linearly between the starting point in West Siberia (ρ_{eff} = 9 persons/km^2) and the end point in Germany (ρ_{eff} = 136 persons/km^2 for PM 10 and 96 persons/km^2 for sulfates and nitrates). This linear interpolation is equivalent to setting the effective population density equal to the mean of the two endpoints of the pipeline. For the pipeline transport within Germany, the reference values for the population exposures were used. This is reasonable since the transport path has a length of 100 kilometers and is therefore likely to sample population densities (within a radius of 100 kilometers) with an average close to the country wide average.

The emission heights for all compressor stations, which only have a minor influence in the above situations, were estimated to be about 50 meters according to photographs shown in (Peebles 1992:138). The population exposures for the emissions from power plant mixes (here: Netherlands and Germany) were set equal to the average values for the respective countries for an emission height of 200 meters. This is a good approximation considering the large number of contributing plants. The ship transport of iron ore was treated in the same way as the ship transport of crude oil from the OECD countries, which will be discussed in section 4.2.2.3.

Results: The results given in tables 4.5 to 4.7 show that the consideration of the spatial differentiation of the population exposures is important in various respects. The population exposures per mass of emitted pollutant (PE/M) for individual processes show a spread of a factor between 10 and about 20 for the three consi-

Table 4.5. Emissions of particulate matter (PM) associated with the supply of 1 TeraJoule of compressed natural gas at a fuel station in Germany: disaggregation according to location and height of emission h

M_{PM} sum [g]	M_{PM} indiv. [g]	loca-tion	h [m]	PE/M [...]	PE [...]	process description / remarks
NL						
extraction 44,1	32,4	NL	200	0,500	1,62 E-2	power plant mix
processing 63,3	29,5	NL	200	0,500	1,48 E-2	power plant mix
pipeline 11,1						all processes < 1 %
D						
extraction 13,2						all processes < 1 %
processing 47,9						all processes < 1 %
NOR						
extraction 20,5						all processes < 1 %
processing 32,6						all processes < 1 %
pipeline 28,7						all processes < 1 %
RUS						
extraction 31,0	13,1	RUS	200	0,027	3,53 E-4	gas-fired power plant
processing 381,5	320,4	RUS	200	0,027	8,64 E-3	power plant mix
	16,6	RUS	50	0,031	5,14 E-4	reduction of iron ore for plant
	14,0	RUS	50	0,031	4,33 E-4	sintering of iron ore for plant
	13,0	RUS	50	0,031	4,02 E-4	burning of cement for plant
pipeline 308,4	121,0	R.-D	50	0,251	3,04 E-2	compressor stations
	73,1	RUS	50	0,031	2,26 E-3	reduction of iron ore for pipeline
	61,9	RUS	50	0,031	1,92 E-3	sintering of iron ore for pipeline
	13,3	RUS	50	0,031	4,12 E-4	steel production for pipeline
	13,2	RUS	200	0,027	3,56 E-4	power plant mix (for processing of natural gas for the compressor)
distrib. D						
pipeline 20,4						all processes < 1 %
compres-sion 270,0	186,8	D	200	0,493	9,21 E-2	power plant mix: electricity for large fuel station
	17,8	RUS	200	0,027	4,80 E-4	power plant mix for processing of natural gas and transport of oil from RUS burnt in power plants
contr. >1%	926,1			0,183	1,69 E-1	73 % of the total mass
contr. <1%	346,7			0,285	9,86 E-2	European average of PE/M
SUM 1273[a]				0,210	2,68 E-1	
SUM small stations 1543[a]				*0,246*	*3,79 E-1*	

indiv. individual process, *PE/M* population exposure per emitted mass [persons (μg/m^3) a/kg], *PE* population exposure [persons (μg/m^3) a], *R..-D* from Russia to Germany, *distrib.* distribution, *contr.* contributions
[a] all PM 10.

Table 4.6. NO_x emissions (as NO_2) associated with the supply of 1 TeraJoule of compressed natural gas at a fuel station in Germany: disaggregation according to location and height of emission h

M_{NOx} sum [g]	M_{NOx} indiv. [g]	loca-tion	h [m]	PE/M [...]	PE [...]	process description / remarks
NL						
extraction 160,2		NL	200			all processes < 1 %
processing 233,3		NL	200			all processes < 1 %
pipeline 266,4	250,5	NL	50	0,167	4,18 E-2	compressor stations
D						
extraction 86,2						all processes < 1 %
processing 437,7	232,3	D	200	0,240	5,58 E-2	power plant mix
NOR						
extraction 244,4	187,5	NOR	200	0,011	2,06 E-3	gas-fired power plant
processing 115,0						all processes < 1 %
pipeline transport 606,3	562,6	NOR	50	0,011	6,19 E-3	compressor stations
RUS						
extraction 572,4	522,5	RUS	200	0,026	1,36 E-2	gas-fired power plant
processing 917,5	318,8	RUS	50	0,026	8,32 E-3	gas burner for process heat
	411,9	RUS	200	0,026	1,08 E-2	power plant mix
pipeline 5242	4840	R.-D	50	0,154	7,44 E-1	compressor stations
distrib. D						
pipeline 434,5	395,9	D	50	0,240	9,50 E-2	compressor stations
compres-sion 4381	3968	D	200	0,240	9,52 E-1	power plant mix D: electricity for large fuel station
contr. >1%	11690			0,162	1,93 E+0	84 % of the total mass
contr. <1%	2307			0,191	4,41 E-1	European average of PE/M
SUM 13997				0,169	2,37 E+0	
SUM small stations 18077				*0,185*	*3,34 E+0*	

indiv. individual process, *PE/M* population exposure to nitrate aerosols per mass of emitted NO_x [persons ($\mu g/m^3$) a/kg], *PE* population exposure to nitrate aerosols [persons ($\mu g/m^3$) a], *R..-D* from Russia to Germany, *distrib.* distribution, *contr.* contributions

dered pollutants. In the sum of the contributions larger than 1 %, these variations in PE/M cancel out to some extent, but not completely. PE/M values for these contributions taken together range between 64 % and 85 % of the European average value used for the remaining processes. This is due to the numerous processes occurring in sparsely populated areas, in particular in Russia. In addition to the effect on the overall results, the spatial differentiation is also associated with

Table 4.7. SO_2 emissions associated with the supply of 1 TeraJoule of compressed natural gas at a fuel station in Germany: disaggregation according to location and height of emission h

M_{SO2} sum [g]	M_{SO2} indiv. [g]	loca-tion	h [m]	PE/M [...]	PE [...]	process description / remarks
NL						
extraction 56,7						all processes < 1 %
processing 81,1						all processes < 1 %
pipeline 11,9						all processes < 1 %
D						
extraction 40,0						all processes < 1 %
processing 163,5	105,6	D	200	0,145	1,53 E-2	power plant mix
NOR						
extraction 27,3						all processes < 1 %
processing 41,9						all processes < 1 %
pipeline 32,3						all processes < 1 %
RUS						
extraction 27,6						all processes < 1 %
processing 1188	1042	RUS	200	0,014	1,41 E-2	power plant mix
	67,4	RUS	50	0,014	9,10 E-4	oil burner for extraction of oil used as fuel for power plant mix
pipeline 316,8	97,4	sea	5	0,016	1,56 E-3	ship transport of iron ore for pipeline
	69,7	RUS	50	0,014	9,41 E-4	sintering of iron ore for pipeline
distribution in D						
pipeline 23,8						all processes < 1 %
compres-sion 2016	1719	D	200	0,145	2,49 E-1	power plant mix D: electricity for large (*small*) fuel station
	80,6	sea	5	0,016	1,29 E-3	ship transport of iron ore for power plant construction
	44,2	RUS	50	0,014	5,97 E-4	extraction of oil for power plants
contr. >1%	3226			0,090	2,84 E-1	80 % of the total mass
contr. <1%	800,0			0,109	8,72 E-2	
SUM 4026				0,092	3,71 E-1	
SUM small 6043 stations				*0,106*	*6,41 E-1*	

indiv. individual process, *PE/M* population exposure to sulfate aerosols per mass of emitted SO_2 [persons ($\mu g/m^3$) a/kg], *PE* population exposure to sulfate aerosols [persons ($\mu g/m^3$) a], *contr.* contributions changes of the ranks of the most significant individual contributions to the overall results. For example, the process with the highest emitted mass of primary particles (part of the processing of natural gas in Russia) only ranks fifth in terms of the associated population exposure.

The information on the location, emission height and associated population exposures of individual processes listed in tables 4.5 to 4.7 will be used in chapter 5 for the purpose of a site-dependent impact assessment with regards to the impact categories of human health and acidification.

4.2.2
Diesel and Petrol

4.2.2.1
Input Data

In 1998, between 75 and 80 % of the automotive fuels Diesel and petrol used in Germany were produced in domestic refineries (MWV 1998:21,26). As a simplification, the direct import of refined fuels to Germany is therefore neglected in the following, i.e. it is assumed that all Diesel and petrol used in Germany is made from raw oil imported to or extracted in Germany. Based on the market shares in 1998, the supply of crude oil in Germany was modeled as a mix of 35 % oil from the OPEC countries and other countries in the Near East and Africa[11], 25 % from Russia, 37 % from the European Union (mainly Norwegian and British fields in the North Sea) and 3 % from domestic extraction (MWV 1998:19-20)[12].

Models of the process chains for the extraction, processing and transport of the crude oil, which differ between the various regions of origin of the raw oil, are contained in the GEMIS database (Rausch et al. 1998; see fig. 4.2). While the oil extracted in the North Sea and in Russia is accordingly brought to Germany via pipelines, the crude oil from the OPEC countries is transported by ship. The processes for the four countries also differ with regards to the method of extraction, which is determined by the degree to which the respective oil fields have been exhausted already: The so-called *primary* extraction, i.e. the direct pumping of the oil to the surface, is only feasible for the first 30 % of an oil field. Subsequently, further oil can be extracted by pumping water into the field. This *secondary* extraction requires a pumping energy which is about three times as high as in the case of the primary extraction (about 0,3 % of the lower calorific value of the oil instead of 0,1 %).

For the oil reservoirs in the OPEC countries and in the North Sea, a share of the primary extraction of 80 % and 50 %, respectively, is assumed. In the case of Russia, no distinction is made between primary and secondary extraction due to a lack of data. The small amounts of oil coming from Germany are produced by

---

[11] The OPEC countries alone contributed 28,2 %. The other countries in Africa, the Near East and other parts of the world, which contribute 6,7 %, are subsumed here under the OPEC countries for the purpose of a site-dependent impact assessment because of their similar geographic locations.

[12] These market shares deviate considerably from the default data in the GEMIS database (Rausch et al. 1998), which are based on a less recent reference (DGMK 1992).

tertiary extraction, where hot water vapor is injected into the fields. This method requires a very high amount of process energy, namely about 25 % of the lower calorific value of the extracted oil (Fritsche et al. 1994:39-40).

Except for these differences, the modeled structure of the process chains is similar for the four countries: The mechanical energy for the extraction is provided either by Diesel engines (OPEC, Russia) or by electrical engines, which are connected to a gas-fuelled power station close to the site of extraction (EU) or to the national grid (D). Subsequent to the extraction, fractions of gas and water are separated off. This requires process heat, which is provided by an oil burner (OPEC, Russia) or a gas burner (EU). For the transport of the oil through pipelines, electrical motors are used, which are supplied with electricity from a gas-fuelled power station (EU) or the national grid (Russia). The ships for the sea transport of oil from the OPEC countries are powered by heavy residual oil which comes from the OPEC countries themselves (Fritsche et al. 1994: 39-43).

In addition to the required amounts of auxiliary energy for the above processes, the burning gas coming from the oil reservoirs also leads to airborne emissions of CO_2. Depending on the country, between 5 and 15 % of the oil gas are not used, but are burned at the location of oil extraction. The resulting CO_2 emissions, which correspond to between 0,3 and 3 % of the CO_2 emissions that would arise from burning the extracted amount of oil itself, are allocated to the process of oil extraction. Furthermore, diffuse losses and the incomplete burning of oil gas lead to emissions of its constituents methane (CH_4) and non-methane volatile organic compounds (NMVOC, Fritsche et al. 1994:41-42). On average over all countries of extraction, the methane losses are about a factor 16 smaller than those arising from the extraction and processing of natural gas and its transport to Germany, while the NMVOC emissions are comparable in both cases (5,8 kg/TJ oil compared to 6,3 kg/TJ natural gas).

In contrast to natural gas, the crude oil imported to Germany is processed at a refinery before the refined products such as petrol or Diesel are distributed to fuel stations. The refining process requires a significant amount of energy, mostly in the form of process heat, which differs between its various co-products. A detailed process chain analysis by Frischknecht et al. (1996:142-144), which allocates energy inputs on the basis of the masses of the outputs in cases of real co-production, concludes that the production of petrol requires four times as much process heat and three times as much electricity as the production of Diesel. Following Fritsche et al. (1994), who include the results of further studies, a factor four for both process heat and electricity demand is adopted here. Accordingly, the demand for process heat (electricity) amounts to 2,75 % (0,25 %) of the lower calorific value of the fuel in the case of Diesel and to 11 % (1,0 %) in the case of petrol (Rausch et al. 1998). The associated Cumulative Energy Demand (CED) for Diesel (4,7 % of H_u, table 4.9) is comparable to the CED for the compression of the natural gas at a large fuel station (5 % of H_u, table 4.4), whereas the associated CED for the refining of petrol is much higher (17,4 % of H_u, table 4.9). Furthermore, the refining process leads to losses of CH_4 and in particular of NMVOC by way of evaporation from tanks, leakages and flaring of certain constituents of the raw oil.

Table 4.8. Relative energy inputs and direct CO_2, CH_4 and NMVOC emissions for the supply of Diesel and petrol at a fuel station in Germany (Rausch et al. 1998)

		raw oil from			
		OPEC	RUS	EU (North Sea)	Germany
market share in Germany [%][a]		35	25	37	3
extraction:		prim. sec.	prim.&sec.	prim. sec.	tertiary
fraction of total extraction	[%]	80 20	100	50 50	100
Diesel	[% of H_u]	0,33 0,9	1,2		
electricity	[% of H_u]			0,1 0,3	0,3
processing:					
process heat	[% of H_u]	0,23	0,3	0,23	25,5
burning of oil gas:					
CO_2 emissions	[kg/TJ]	730	2384	260	260
CH_4 emissions	[kg/TJ]	5,2	27,5	2,2	2,0
NMVOC emissions	[kg/TJ]	3,0	17,0	1,2	3,0
transport to Germany:					
means of transportation		ship	pipeline	pipeline	/
transport distance	[km]	8800	2500	500	/
energy carrier		heavy oil	electricity	electricity	/
transport energy	[% of H_u]	2,2	0,4375	0,0875	/

refining:		Diesel	petrol
electricity	[% of H_u]	0,25	1,0
process heat	[% of H_u]	2,75	11,0
CH_4 emissions	[kg/TJ]	1,0	1,0
NMVOC emissions	[kg/TJ]	10,0	10,0
regional distribution in Germany:			
means of transportation		truck	
transport distance	[km]	100	
energy carrier		Diesel	
transport energy	[% of H_u]	0,28	
fuel station:		Diesel	petrol
electricity	[% of H_u]	0,01	0,01
NMVOC emissions	[kg/TJ]	0	133[b]

H_u lower calorific value, *prim.* primary, *sec.* secondary
[a] in 1998 (MWV 1998:19-20).
[b] Besides the emissions from tanks and from the fueling process at the fuel station, this figure also comprises the evaporative losses from the vehicles. It represents a medium value between the actual emissions in 1990 (295 kg/TJ) and possible future reductions (down to 36 kg/TJ), as they were determined for the German state of Baden-Württemberg (Krüger et al. 1997:11, 21; production and consumption data from Hedden et al. 1994:4-14).

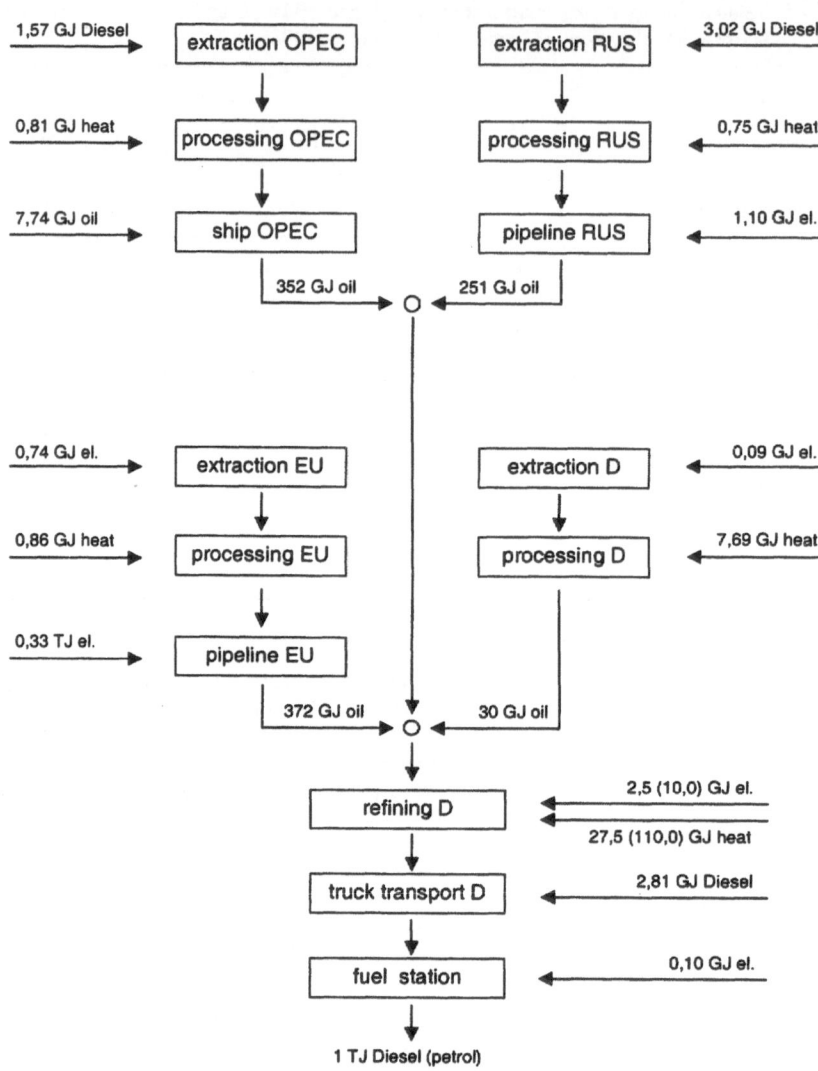

Fig. 4.2. Main energy flows associated with the supply of Diesel or petrol (numbers in brackets for the refining stage) at a fuel station in Germany according to the data in table 4.8. *el.* electricity

From the refinery, the Diesel or petrol is brought to fuel stations by a Diesel-fueled truck, for which a transport distance of 100 kilometers was assumed (Rausch et al 1998). The electricity demand of the fuel station itself amounts to 0,1 % of the lower calorific value of the Diesel or petrol. In contrast to the high energy demand for the compression of natural gas at a fuelling station, it therefore does not play a significant role. In the case of petrol, the fueling process as well as evaporations from tanks at the fuel stations lead to high NMVOC emissions. The figure of 133 kg NMVOC/TJ petrol used here also comprises the evaporative losses from the vehicles. It represents a medium value between the actual emissions in 1990 (295 kg/TJ) and possible future reductions (down to 36 kg/TJ), as they were determined for the German state of Baden-Württemberg (Krüger et al. 1997:11, 21; production and consumption data from Hedden et al. 1994:4-14).

The amounts of auxiliary energy required for each of the processes described above are shown in table 4.8 together with the direct emissions of CO_2, CH_4 and NMVOC. The structure of the process chains associated with the supply of Diesel or petrol at a gas station in Germany is shown in fig. 4.2. The absolute values of the energy flows, which result from combining the relative values given in table 4.7 with an amount of 1 TJ of Diesel or petrol supplied at the fuel station, are also indicated.

4.2.2.2
Emissions and Cumulative Energy Demand

Based on the input data given in table 4.8, the Cumulative Energy Demand (CED) and the cumulated emissions of CO_2, CH_4, N_2O, PM, SO_2, NO_x and NMVOC associated with the supply of 1 TJ of Diesel fuel or petrol at a fuel station in Germany were calculated using the GEMIS software. The results are shown in the last row of table 4.9. The table also contains the emissions and energy demands associated with each phase (extraction, transport, refining, distribution) of the fuel supply chain. These contributions to the overall sum were calculated by defining separate processes within the GEMIS software[13]. Since the Cumulative Energy Demand (CED) also contains the lower calorific value H_u of the supplied fuel itself, table 4.9 lists the differences CED-H_u (in GJ as well as in % relative to H_u) as a more reasonable measure of the energy demand associated with the supply of the fuel, which does not include the burning of the fuel.

The refining process is the phase with the highest value of CED-H_u in absolute terms and, except for the insignificant extraction in Germany, also in relative terms (4,7 % of H_u for Diesel and 17,4 % for petrol). This is also reflected in the associated high emissions. However, in the case of Diesel, the emissions of some pollutants (PM, SO_2 and NO_x) from the ship transport of OPEC oil are significantly higher despite the lower energy demand. This is due to the higher emission factors of the ocean transport ships compared to the burner mix that is used to provide the heat for the refining processes. Per GigaJoule of fuel input, the ship emits 984 g SO_2, 1000 g NO_x and 100 g PM, compared to 98 g SO_2, 73 g NO_x and 2 g PM for the mix of burners (20 % gas, 40 % oil and 40 % propane) at the refinery. For the same reason, the emissions of particulate matter and methane through the extraction of oil in Russia are considerably higher than those from the refining of Diesel in Germany despite the lower energy demand.

However, the comparison between the emissions from the refinery and the emissions from the ship is a good example to demonstrate the need for a site-dependent impact assessment of emissions within a Life Cycle Inventory. Both with regards to human health impacts and with regards to acidification it is qualitatively

---

[13] see footnote 7 above.

Table 4.9. Emissions and Cumulative Energy Demand (*CED*) associated with the supply of 1 TeraJoule (lower calorific value H_u) of Diesel or petrol at a fuel station in Germany - **Part 1:** Pollutants with regional range of dispersion

	PM [g]	SO$_2$ [g]	NO$_x$ [g]	NMVOC [g]
raw oil from D (H_u=30,2 GJ)				
extraction[a]	3,7	312,9	532,2	142,8
raw oil from EU (H_u=371,9 GJ)				
extraction[a]	15,9	22,8	668,8	469,6
pipeline transport	10,0	14,9	271,2	9,3
raw oil from RUS (H_u=251,3 GJ)				
extraction[a]	617,2	4706,7	3787,7	4464,3
pipeline transport	562,0	1912,9	761,4	43,9
raw oil from OPEC (H_u=351,8 GJ)				
extraction[a]	197,4	2901,3	1793,1	1103,7
transport by ship	853,7	8891,0	8102,6	356,2
refining in D (H_u=1,0 TJ)				
Diesel	288,3	4123,2	3670,1	10505,9
(petrol)	(1100,8)	(16421,6)	(14539,4)	(12017,6)
distribution in D (1,0 TJ)				
transport by truck	140,6	176,2	2697,0	375,3
fuel station Diesel	1,6	11,8	25,7	2,6
(fuel station petrol)	(1,6)	(11,8)	(25,7)	(133000,0)
SUM Diesel	2690,2	23073,8	22309,7	17473,6
(SUM petrol)	(3502,7)	(35372,2)	(33179,0)	(151982,8)

[a] The extraction phase also comprises the processing step and the direct emissions from the burning of oil gas. This is because these three steps are modeled as one process in the GEMIS software, in contrast to the case of natural gas, where extraction and processing are modeled as two separate processes (see table 4.3).

plausible that the emissions from the ships should be considered less important because the population density as well as the susceptibility for acidification in some area around the ships are zero. With regards to human health impacts, a similar reasoning applies to emissions occurring in sparsely populated areas such as the oil fields in Russia. In the following section and in chapter 5, this qualitative argument will be put in quantitative terms on the basis of the population exposures per mass of pollutant derived in chapter 3.

Table 4.9. Emissions and Cumulative Energy Demand (*CED*) associated with the supply of 1 TeraJoule (lower calorific value H_u) of Diesel or petrol at a fuel station in Germany - **Part 2:** Pollutants with hemispherical or global range of dispersion and CED

	CO_2 [kg]	CH_4 [g]	N_2O [g]	CED-H_u [GJ]	CED/H_u-1 [%]
raw oil from-D (H_u=30,2 GJ)					
extraction[a]	569,6	144,5	13,3	9,3	30,9
raw oil from EU (H_u=371,9 GJ)					
extraction[a]	300,6	1072,7	7,1	3,5	0,9
pipeline transport	65,3	101,7	2,6	1,1	0,3
raw oil from RUS (H_u=251,3 GJ)					
extraction[a]	1002,0	7130,4	13,6	5,3	2,1
pipeline transport	310,4	501,8	10,9	4,8	1,9
raw oil from OPEC (H_u=351,8 GJ)					
extraction[a]	497,0	1898,2	7,8	3,2	0,9
transport by ship	726,9	173,4	5,2	9,7	2,8
refining in D (H_u=1,0 TJ)					
Diesel	2678,0	4534,9	47,4	47,4	4,7
(petrol)	(10650,0)	(14906,1)	(188,6)	(173,9)	(17,4)
distribution in D (1,0 TJ)					
transport by truck	252,7	100,3	18,1	3,4	0,3
fuel station Diesel	18,0	42,9	0,7	0,3	0,0
(fuel station petrol)	(18,0)	(42,9)	(0,7)	(0,3)	(0,0)
SUM Diesel	6420,5	15700,8	126,6	88,0	8,8
(SUM petrol)	(14392,5)	(26072,0)	(267,8)	(214,6)	(21,4)

[a] The extraction phase comprises the processing step and the direct emissions from the burning of oil gas. This is because these three steps are modeled as one process in the GEMIS software, in contrast to the case of natural gas, where extraction and processing are modeled as two separate processes (see table 4.3).

4.2.2.3
Spatial Disaggregation

Input data: In order to combine the emission inventory contained in table 4.9 with the population exposures derived in chapter 3 for the pollutants PM, SO_2 and NO_x, the inventory for these pollutants was further disaggregated according to the different locations and heights of the contributing processes. According to the cutoff criterion introduced in section 4.2.1.3, the disaggregation was only carried out for those processes that contribute more than 1 % to the overall cumulated emission for a particular pollutant.

Tables 4.10 through 4.12 show the disaggregated emission inventories for the pollutants PM, SO_2 and NO_x for the provision of 1 TeraJoule of Diesel at a fuel station in Germany. The inventories for petrol only differ with regards to the emissions from the refining phase, which are roughly four times as high as in the case of Diesel. The corresponding final results are shown in table 4.13 for all three pollutants together. In each of the tables 4.10 through 4.12, the first column contains the emissions associated with the various life cycle stages as given in table 4.9, part 1. The columns 2 to 7 contain information about the individual processes within these stages that contribute more than 1 % to the sum over the whole fuel chain. For the Diesel and petrol fuel chain, these values add up to between 91 and 96 % of the total value.

For the most part, the processes contributing more than 1 % to the overall emissions are those supplying the energy for the extraction, processing, transport, refining and distribution of the fuels according to the energy demands listed in table 4.8. In some cases, processes upstream of the respective power plants, motors or burners also contribute significantly. This is the case for the PM, SO_2 and NO_x emissions from extracting, processing and refining the oil that fuels the ships transporting raw oil from the OPEC countries to Germany (tables 4.10 to 4.12). It also applies to the NO_x (for petrol also the PM and SO_2) emissions from the ships transporting oil from OPEC countries to Germany where it is used in oil burners during the refining of raw oil (table 4.11). For petrol, this is furthermore the case for the PM emissions from the supply of the oil from Russia that is used in the oil burners of the refinery in Germany. Contrary to the case of natural gas, where some processes of materials production for the facilities of the fuel chain contributed significantly to the overall emissions, all of these processes remain below the 1 % threshold here.

With regards to the emission heights of the individual processes listed in tables 4.10 to 4.12, the standard assumptions that were made in the case of natural gas (tables 4.5 to 4.7) also apply here. In terms of the location of the processes and the corresponding population exposures or effective population densities, power plant mixes and pipeline transports were treated in the same way as in the case of natural gas, and the same effective population densities were used for processes in Russia.

Processes not covered by these analogies are discussed in more detail in the following. The population exposures associated with emissions from ships transporting crude oil from the OPEC countries to Germany depend on the transport route. In 1997, about 50 % of the OPEC oil in Germany was imported from Libya and Algeria (European Oil and Gas Yearbook 1999:2-25). The tanker route for these imports was assumed to lead through the Mediterranean, the street of Gibraltar and the North Eastern Atlantic to Rotterdam (about 4000 kilometers). Population exposures for points along this route were determined with EcoSense, resulting in averages of 0,446, 0,085 and 0,112 persons ($\mu g/m^3$) a/kg for Diesel soot particles, sulfate and nitrate aerosols respectively. This corresponds to about 40 %, 59 % and 46 % of the respective German averages for traffic emissions. Hence, the population exposures from the ship emissions on this route cannot be neglected. For sea transports occurring close to populated land masses (e.g. in the

Table 4.10. PM emissions associated with the supply of 1 TeraJoule of Diesel at a fuel station in Germany: disaggregation according to location and height of emission h

M_{PM} sum [g]	M_{PM} indiv. [g]	loca-tion	h [m]	PE/M [...]	PE [...]	process description / remarks
D						
extraction 3,7						all processes < 1 %
EU						
extraction[a] 15,9						all processes < 1 %
pipeline 10,0						all processes < 1 %
RUS						
extraction[a] 617,2	499,0	RUS	5	0,197	9,85 E-2	Diesel engine for extraction
	66,9	RUS	50	0,031	2,07 E-3	oil burner for processing
	45,6	RUS	200	0,027	1,23 E-3	power plant mix
pipeline 562,0	547,8	R.-D	200	0,219	1,20 E-1	power plant mix
OPEC						
extraction[a] 197,4	132,2	OPEC	5	0,724	9,57 E-2	Diesel engine for extraction
	56,5	OPEC	50	0,113	6,41 E-3	oil burner for processing
shipping 853,7	774,9	sea	5	0,083	6,43 E-2	sea transport
	47,0	OPEC	50	0,113	5,33 E-3	oil burner for processing and refining of oil fueling the ship
refining D						
refining 288,3	157,8	D	50	0,705	1,11 E-1	oil burner for 20% process heat
distrib. D						
transport by truck 140,6	69,7	D (II,1)	5	1,725	1,20 E-1	Diesel truck, 50% within cities
	52,5	D (II,3)	5	1,439	7,55 E-2	Diesel truck, 50% outside of cities
fuel station 1,6						all processes < 1 %
contr. >1% 2450				0,286	7,00 E-1	91 % of the total mass
contr. <1% 240				0,285	6,84 E-2	European average for PE/M
SUM 2690[b]				0,286	7,69 E-1[c]	

indiv. individual process, *PE/M* population exposure per emitted mass [persons (μg/m^3) a/kg], *PE* population exposure [persons (μg/m^3) a], *R.-D* from Russia to Germany, *distrib.* distribution, *contr.* contributions

[a] includes processing.

[b] sum of 1528 g Diesel particles (PM 2,5) and 1162 g other particles (PM 10).

[c] sum of 4,54 E-1 persons (μg/m^3) a exposure to Diesel particles (PM 2,5) and 3,14 E-1 persons (μg/m^3) a exposure to other particles (PM 10).

Table 4.11. NO_x emissions associated with the supply of 1 TeraJoule of Diesel at a fuel station in Germany: disaggregation according to location and height of emission h

M_{NOx} sum [g]	M_{NOx} indiv. [g]	loca-tion	h [m]	PE/M [...]	PE [...]	process description / remarks
D						
extraction 532,2	506,9				1,22 E-1	all processes < 1 %
EU						
extraction[a] 668,8	576,9	NOR	200	0,011	6,35 E-3	gas-fueled power plant
pipeline 271,2	252,0				2,77 E-3	all processes < 1 %
RUS						
extraction[a] 3788	3493	RUS	5	0,026	9,12 E-2	Diesel engine for extraction
	223,1	RUS	50	0,026	5,82 E-3	oil burner for processing
pipeline 761,4	700,3	R.-D	200	0,154	1,08 E-1	power plant mix
OPEC						
extraction[a] 1793	1586	OPEC	5	0,096	1,52 E-1	Diesel engine for extraction
	188,4	OPEC	50	0,096	1,80 E-2	oil burner for processing
shipping 8103	7749	sea	5	0,021	1,63 E-1	sea transport
	173,1	OPEC	5	0,096	1,66 E-2	Diesel engine for extraction and refining of oil fueling the ship
	156,8	OPEC	50	0,096	1,50 E-2	oil burner for processing and refining of oil fueling the ship
refining D						
refining 3670	2438	D	50	0,240	5,85 E-1	burner mix for process heat (oil 20%, gas 40%, propane 40%)
	599,9	D	200	0,240	1,44 E-1	power plant mix
	273,1	sea	5	0,021	5,74 E-3	sea transport of OPEC oil to oil burners for process heat
distrib. D						
transport by truck 2697	1516	D (II,1)	5	0,240	3,64 E-1	Diesel truck, 50% within cities
	1084	D (II,3)	5	0,240	2,60 E-1	Diesel truck, 50% outside of cities
fuel station 25,7						all processes < 1 %
contr. >1%	21517			0,096	2,06 E+0	96 % of the total mass
contr. <1%	792,8			0,191	1,51 E-1	European average for PE / M
SUM 22310				0,099	2,21 E+0	

indiv. individual process, *PE/M* population exposure to nitrate aerosols per mass of emitted NO_x [persons $(\mu g/m^3)$ a/kg], *PE* population exposure to nitrate aerosols [persons $(\mu g/m^3)$ a], *R.-D* from Russia to Germany, *distrib.* distribution, *contr.* contributions
[a] includes processing.

Table 4.12. SO_2 emissions associated with the supply of 1 TeraJoule of Diesel at a fuel station in Germany: disaggregation according to location and height of emission h

	M_{SO2} sum [g]	M_{SO2} indiv. [g]	loca-tion	h [m]	PE / M [...]	PE [...]	process description / remarks
D							
extraction	312,9	300,8	D	50	0,145	4,36 E-2	oil gas burner for processing
EU							
extraction[a]	22,8						all processes < 1 %
pipeline	14,9						all processes < 1 %
RUS							
extraction[a]	4707	3027	RUS	5	0,014	4,09 E-2	Diesel engine for extraction
		1523	RUS	50	0,014	2,06 E-2	oil burner for processing
pipeline	1913	1767	R.-D	200	0,080	1,40 E-1	power plant mix
OPEC							
extraction[a]	2901	1604	OPEC	5	0,050	7,94 E-2	Diesel engine for extraction
		1286	OPEC	50	0,050	6,37 E-2	oil burner for processing
shipping	8891	7631	sea	5	0,016	1,22 E-1	sea transport
		1070	OPEC	50	0,050	5,30 E-2	oil burner for processing and refining of oil fueling the ship
		175,0	OPEC	5	0,050	8,66 E-3	Diesel engine for extraction and refining of oil fueling the ship
refining D							
refining	4123	3261	D	50	0,145	4,73 E-1	oil burner for 20% process heat
distrib. D							
transport by truck	176,2						all processes < 1 %
fuel station	11,8						all processes < 1 %
contr. >1%		21644			0,048	1,05 E+0	94 % of the total mass
contr. <1%		1430			0,109	1,56 E-1	European average for PE/M
SUM	23074				0,052	1,20 E+0	

indiv. individual process, *PE/M* population exposure to sulfate aerosols per mass of emitted SO_2 [persons ($\mu g/m^3$) a/kg], *PE* population exposure to sulfate aerosols [persons ($\mu g/m^3$) a], *R.*-D from Russia to Germany, *distrib.* distribution, *contr.* contributions
[a] includes processing.

Mediterranean sea), the complete discarding of emissions from ships, which is sometimes suggested, is therefore not a good approximation.

A further 25 % share of the OPEC oil is imported from the Near East mainly on a shipping route of about 25000 kilometers leading around the southern tip of

Table 4.13. Total population exposures to PM, sulfate and nitrate aerosols due to emissions of PM, SO_2 and NO_x, respectively, from the supply of 1 TeraJoule of petrol at a fuel station in Germany

	M sum [g]	M indiv. [g]	PE/M [...]	PE [...]	process description / remarks
PM					
contr. >1%		3259	0,336	1,10 E+0	93 % of the total mass
contr. <1%		244	0,285	6,93 E-2	European average for PE/M
SUM	3503[a]		0,333	1,17 E+0[b]	
NO_x					
contr. >1%		31388	0,136	4,25 E+0	95 % of the total mass
contr. <1%		1792	0,191	3,42 E-1	European average for PE / M
SUM	33179		0,139	4,60 E+0	
SO_2					
contr. >1%		33502	0,078	2,63 E+0	95 % of the total mass
contr. <1%		1871	0,109	2,04 E-1	European average for PE/M
SUM	35372		0,080	2,83 E+0	

indiv. individual process, *PE/M* population exposure per mass [persons ($\mu g/m^3$) a/kg], *PE* population exposure [persons ($\mu g/m^3$) a]
[a] sum of 1679 g Diesel particles (PM 2,5) and 1824 g PM 10.
[b] sum of 4,72 E-1 persons ($\mu g/m^3$) a exposure to Diesel particles (PM 2,5) and 6,93 E-1 persons ($\mu g/m^3$) a exposure to PM 10.

Africa. The remaining 25 % of the OPEC imports to Germany come from Nigeria and Venezuela with a shipping distance of about 10000 kilometers. In all cases, the shipping routes are typically at least 1000 kilometers away from the shores, and the closest land masses (Africa) have a very low average population density. Therefore, the population exposures can be set to zero in good approximation in these cases.

The GEMIS data on the ship emissions are based on an average transport distance of 8800 kilometers, which is consistent with weighting the above three transport distances with the respective percentages of OPEC oil transported along these paths. Weighting the respective population exposures for the individual routes in the same way leads to the population exposures per mass of emitted pollutant listed in tables 4.10 to 4.12.

The effects of emissions from refineries were modeled for the German refinery mix. In 1997, 14 refineries were operating in Germany. While airborne emissions in some refineries partly occur through very high stacks, the typical emission height may rather be in the order of 50 meters (IFEU 1999), except for the diffuse emissions of NMVOC occurring closer to the ground. The locations and capacities of the German refineries (Shell 1998a:15) and their distribution across the nine settlement structure classes in Germany in table 4.14. About 50% of the refining capacity is located in settlement structure classes with above average population

Table 4.14. Geographical distribution of oil refineries in Germany (Shell 1998a:15), expressed in terms of settlement structure classes (notation as defined in table 3.3)

class	I,1	I,2	II,1	I,3	II,3	I,4/5	II,4/5	III,4	III,5	sum
number of refineries	4	2	1	0	2	0	1	2	2	14
capacity [mio. t/a]	34,4	14,5	5,0	0,0	19,0	0,0	12,0	7,2	16,0	108,1
capacity [%]	32	13	5	0	18	0	11	7	15	100

exposures, namely (I,1), (I,2) and (II,1). The relative distribution of the capacity was used to weight the population exposures associated with the settlement structure classes. The emissions of particulate matter (PM 10) from oil burners, for which a release height of 50 meters was assumed, are accordingly associated with a population exposure of 0,8 persons (μg/m^3) a/kg, which is 1,35 times the country average value for that emission height.

The distribution of the Diesel or petrol from the refineries to the fuel stations by truck was assumed to occur to 50 % within cities, which were represented here by the settlement structure class (II,1), and to 50 % outside of cities, which was represented by the ‚average' class (II,3) (Rausch et al. 1998)[14]. This results in a overall population exposure to Diesel particles of 1,45 persons (μg/m^3) a/kg, which is 1,14 times the country average value for traffic emissions in Germany.

Finally, the impacts from emissions of the oil extraction from the North Sea were modeled for Norwegian oil fields, such that the same data for the population exposures were used as in the case of natural gas. The impacts for the emissions from the British oil fields can be assumed to be similar due to their geographic proximity.

Results: As in the case of natural gas, the results presented in tables 4.10 to 4.12 show that the consideration of the spatial differentiation of the population exposures is significant in various respects. The specific population exposures per mass of emitted pollutant (PE/M) for individual processes show a spread of a factor between 10 and about 50 for the three considered pollutants, with the large factor of 50 applying to the spread between traffic emissions of Diesel soot particles in Germany and power plant emissions of particles in Russia. In the sum of the contributions larger than 1 %, these variations in PE/M cancel out to some extent, but not completely. In the case of Diesel, PE/M values for these contributions taken together range between 44 % and 100 % of the European average value used for the remaining processes. This is due to the numerous processes occurring in sparsely populated areas, in particular at sea, in Russia and in the OPEC countries. In the case of petrol, the total site-dependent PE/M values for the contributions larger than 1 % of the total emissions range between 71 % and 118 % of the European averages due to the larger share of the refineries in Germany.

---

[14] The percentages refer to the transport distances, while the emissions are somewhat higher for urban driving.

In addition to the effect on the overall results, the spatial differentiation is also associated with changes of the ranks of the most significant individual contributions to the overall results. In particular, the process with the by far highest emitted masses of particles, SO_2 and NO_x in the case of Diesel, namely the ocean shipping of OPEC oil, only ranks fifth, fourth and third, respectively, in terms of the associated population exposures.

The information on the location, emission height and associated population exposures of individual processes listed in tables 4.10 to 4.12 will be used in chapter 5 for the purpose of a site-dependent impact assessment with regards to the impact categories of human health and acidification.

4.2.3
Comparison of the Upstream Emissions

Table 4.15 and fig. 4.3 summarize the emissions and the Cumulative Energy Demands for the supply of 1 TeraJoule of Diesel, petrol and natural gas, respectively, at a fuel station in Germany. With regards to the buses, only the comparison between natural gas and Diesel is of interest. In the reference case of a large fuel station for natural gas, the Cumulative Energy Demand (less the lower calorific value H_u of the provided fuel) is somewhat higher than for Diesel. However, the emissions of CO_2, PM, SO_2 and NO_x are smaller for natural gas since many processes of the natural gas fuel chain are powered by natural gas rather than oil as in the case of the Diesel fuel chain and are therefore associated with lower emission factors. In particular, the emissions of PM, SO_2 and NO_x for the ocean ships used for the transport of raw oil from the OPEC countries to Germany are very high, such that they represent the largest contributions to the overall emissions of these

Table 4.15. Comparison of the emissions [kg/TJ] and the Cumulative Energy Demands (*CED*) for the supply of 1 TeraJoule (lower calorific value H_u) of Diesel, petrol and compressed natural gas at a fuel station in Germany

	Diesel	petrol	compressed natural gas	
			large station	small station
(CED-H_u)/H_u [%]	8,8	21,4	9,4	14,4
CO_2	6421	14393	5427	8497
CH_4	15,7	26,1	170,4	177,7
N_2O	0,1	0,3	0,2	0,3
CO_2-equivalents[a]	6789	15023	9067	12327
PM 10	1,2	1,8	1,3	1,5
PM 2,5	1,5	1,7	0,0	0,0
NO_x (as NO_2)	22,3	33,2	13,7	18,1
SO_2	23,1	35,4	4,0	6,0
NMVOC	17,5	152,0	7,4	7,8

[a] according to the Global Warming Potentials integrated over 100 years from (Houghton et al. 1995; see table 2.1). Strictly speaking, the calculation of CO_2-equivalents is part of the impact assessment phase. It is nevertheless added here and in the following tables of chapter 4 to facilitate an intermediate comparison of the emissions.

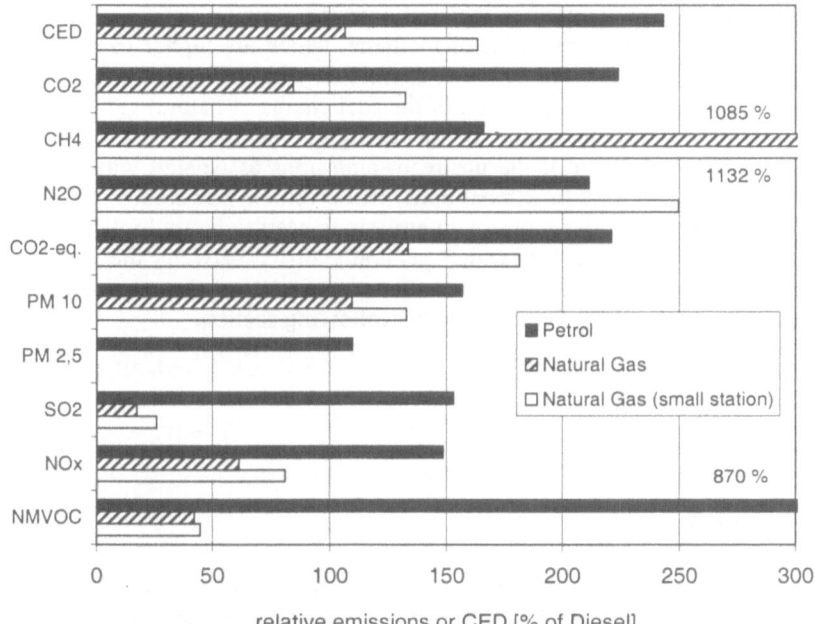

Fig. 4.3. Comparison of the emissions and the Cumulative Energy Demands (*CED-H$_u$*) for the fuel supply chains of petrol and natural gas (large and small fuel station) relative to those of Diesel, based on the supply of 1 TeraJoule (lower calorific value H_u) of fuel at a fuel station in Germany

pollutants in the supply chain of Diesel (see tables 4.10 to 4.12). The Cumulative Energy Demand can therefore not be used as a proxy indicator for the emissions of energy-related airborne pollutants here.

Even though the CO_2 emissions for the natural gas fuel supply chain are lower in the reference case of a large fuel station, this advantage is overcompensated with regards to the CO_2-equivalents by the methane emissions, which are about 10 times as high as those for the Diesel fuel chain and mainly result from losses of natural gas rather than from combustion processes. The latter also applies to the NMVOC emissions in both cases. Contrary to the methane emissions, they are higher in the case of Diesel, which results from their higher share in oil gas compared to natural gas and from their high emissions in refineries.

The doubling of the energy required for the compression of natural gas at a small fuel station compared to a large station notably increases all emissions and the energy demand. In particular, the disadvantages of natural gas in terms of the CO_2-equivalents and the CED strongly increase, while the advantages with regards to PM, SO_2, NO_x and NMVOC become somewhat smaller. This underscores the importance of connecting natural gas fuel stations to pipelines providing high input pressures in the order of 10 bar in order to minimize the energy required for the compression of the gas.

The supply of petrol is associated with by far the highest CED and emissions due to the high energy demand associated with the refining process. This holds with the exception of the emissions of methane, which are higher in the case of natural gas mainly due to the large losses associated with the import from Russia.

So far, the basis for the comparison of the fuel chains of natural gas and Diesel was the supply of the same amount of fuel energy (1 TeraJoule) at a fuel station. However, for a comparison of the upstream emissions associated with driving a distance of 1 kilometer (the functional unit) with a vehicle powered by either Diesel, petrol or natural gas, these upstream emissions need to be multiplied with the fuel consumptions per distance, which generally differ between the considered vehicles. The fuel consumptions and use phase emissions of the considered vehicles (buses and cars) will be described in the following section.

4.3
Buses

4.3.1
Vehicle Operation

There is a large number of references on emissions and fuel consumptions of Diesel buses in Germany. In particular, the „Handbook on Emission Factors of Road Vehicles" (INFRAS 1999) offers a consistent database for a large variety of vehicle types and serves as a standard reference. The dynamic emission factors for commercial vehicles in this database refer to representative driving patterns which are differentiated according to vehicle type, street type, longitudinal slope, traffic situation and the load of the vehicle. They are based on emission measurements for a series of representative motors according to two static measurement protocols (ECE 13-step test and an extended measurement at 35 load points) and two dynamic test cycles (US-transient cycle and FIGE-cycle). On this empirical basis, emission factors are derived through computer simulation of the vehicle dynamics for a set of driving patterns representing various traffic situations on different street types (Hassel et al. 1995).

The database on fuel consumptions and emissions of natural gas vehicles in general is less solid at present. This also applies to natural gas buses, even though there tend to be more data available on buses than on other NGVs such as passenger cars, taxis, garbage collection trucks or delivery trucks. Some projects to determine more reliable emission factors and fuel consumptions of natural gas vehicles are underway, in particular the NGVEurope project (see www.snafu.de/~innotec/NGVeurope) and the demonstration program of the German Federal Environmental Agency (Rodt et al. 1998), but their full results were not available at the time of writing[15].

---

[15] The results of a measurement program for various alternative fuel vehicles in the United States, funded by the Department of Energy, are available at www.ott.doe.gov/ohvt/ heavy_vehicle/ hv/emisbus.html. However, most of the measurements date back to around 1995. Since significant technical improvements of NGVs with regards to motor management and catalytic converters can be expected to have occurred since then (see the discussion of the measurement series on buses in Brussels below), the respective data for natural gas buses are not used here.

Table 4.16. Selection of fuel consumptions and dynamic emission factors for natural gas buses. Reference values for Diesel buses are indicated in brackets where available.

	fuel [MJ/km]	PM [g/km]	NO$_x$ [g/km]	CH$_4$ [g/km]	NMVOC [g/km]
Brussels (quoted in Keller and Kessler 1998:46)					
(1) 1994, no catalytic converter			10,5	2,835[a]	0,315[a]
(2) 1995, motor management and special catalytic converter			4	0,108[a]	0,012[a]
(3) 1995, regular catalytic converter			4,5	0,315[a]	0,035[a]
(4) 1996, regular catalytic converter			9,5	4,05[a]	0,45[a]
(5) 1996, special catalytic converter			4,5	1,215[a]	0,135[a]
(6) (Diesel EURO 1 reference)			(19,5)		(1[b])
Rüsselsheim (Rausch et al. 1998)	21,95	0,127	6,35	1,02	0,254
Basel (Keller and Kessler 1998:48)	19,6		5,58	2,19[a]	0,243[a]
(Diesel EURO2 reference)	(15,42)			(0,021)	(0,821)
Dublin, lean burn engine (NGV-Europe project[c])	24,54		7,2	1,74	0,3
(Diesel reference)	(19,29)		(15,68)	(0)	(0,85)
empirically based analogy to smaller vehicles (Mangelsdorf et al. 1999:23)		0,005	1,0		
(Diesel EURO 2 reference)		(0,24)	(11)		

[a] Measured total VOC emissions were assumed to consist of 90 % CH$_4$ and 10 % NMVOC, which is a general rule of thumb for NGVs.
[b] Measured total VOC emissions were assumed to be 100 % NMVOC.
[c] see www.snafu.de/~innotec/NGVeurope.

Therefore, an overview of literature data on fuel consumptions and emissions of buses fueled by natural gas will be provided. Rather than claiming to be complete in any way, the focus of the overview is on more recent data, since significant improvements can be expected to occur in the current early phase of technological development of NGVs. Furthermore, the overview will be limited to emission factors which refer to dynamic driving conditions of vehicles and are given in units of pollutant mass per distance of driving. Stationary emission factors for engines alone, on the other hand, such as they are determined e.g. in the European 13-mode certification test, are given in units of pollutant mass per output work of the motor (e.g. g/kWh). Since their conversion into distance-related emission factors is not straightforward, they will not be considered here.

A selection of dynamic fuel consumption and emission factors is provided in tables 4.16 (emissions and fuel consumptions) and 4.17 (fuel consumptions only). The results of reference measurements on Diesel buses are also indicated where available. Most emission factors result directly from measurements on the vehi-

Table 4.17. Selection of fuel consumptions of natural gas buses. Reference values for Diesel buses are indicated, where available, together with the relative difference Δ = (natural gas-Diesel) / Diesel of the fuel consumptions.

	natural gas [MJ/km]	Diesel [MJ/km]	Δ [%]
Berlin (Rösgen et al. 1997), long term means			
(1) 4 standard buses	18,4		
(2) 4 articulated buses, MAN	21,5		
(3) 2 articulated buses, Mercedes	24,3		
Berlin, direct comparisons			
(1) standard bus	20,7	15,7	+32
(2) standard bus	13,2	12,1	+9
(3) standard bus	16,8	14,0	+21
(4) articulated bus	24,5	21,7	+13
(5) articulated bus	22,3	18,6	+20
(6) articulated bus	26,0	18,6	+40
Basel (Keller and Kessler 1998:48)	19,6	15,4	+27
Augsburg (NGV-Europe project[a])			
(1) standard bus, heating included	17,2	12,1	+42
(2) articulated bus, heating included	22,0	18,3	+20
Dublin, lean burn engine (NGV-Europe project[a])	24,5	19,3	+27
other NGV-Europe cities (Bemtgen et al. 1997)			
Hannover, standard bus	20,5		
Hannover, articulated bus	22,1		
Apeldoorn, 2 standard buses, mean	20,6		
Saarbrücken, standard and articulated	18,8		
minimum / mean / maximum Δ			+10/25/40

[a] see www.snafu.de/~innotec/NGVeurope.

cles, either on test stands or on the road. One reference (Mangelsdorf et al. 1999) combines results of measurements carried out for smaller vehicles with an analogy reasoning to transfer them to buses. In each case, the emission factors and fuel consumptions refer to a particular driving cycle represented by the vehicle velocity v(t) as a function of time. This may either be a synthetic test cycle or a cycle recorded in actual traffic. Given the sparse data on NGV emissions, the mixing (e.g. averaging) of emission factors based on different methods and driving cycles cannot be avoided.

Comparing the emission factors for Diesel buses and natural gas buses in table 4.16 shows that the aim of the introduction of natural gas vehicles of reducing the emissions of regulated pollutants can be most clearly fulfilled in the case of particulate matter (PM). The degree of emission reduction (if any) for the other pollutants depends much more on a technical optimization that cannot be taken for granted, in particular for the earlier exemplars of NGVs. The series of measurements in Brussels showed that the motor management needs to be specifically adjusted to the new fuel. Furthermore, special catalytic converters may be required for natural gas, which may also have to be exchanged after a certain operating period (Keller and Kessler 1998:46). In that sense, the lower emission values in table 4.16 indicate upper limits for the emissions that can be achieved by proper technical optimization.

To some extent, this also applies to the fuel consumptions, which are between 10 % and 40 % higher for natural gas buses than for Diesel buses, if the values for the same cities are compared (table 4.17). The average increase of the fuel consumption of 25 % will be used as a reference value for all of the following calculations, with the range of 10-40 % being used for sensitivity analyses. The reasons for the increased fuel consumption of the natural gas buses are

- the sometimes still deficient technical optimization of the natural gas engines in relation to the regionally varying gas qualities, compared to the mature Diesel technology
- the additional mass of the pressurized gas tanks, which depends on the materials used and the intended range of the buses and is to some extent inevitable. For buses with a range of about 280 kilometers and composite steel tanks wrapped with aramid fibres, the additional mass amounts to about 13 % of the total mass of the buses (Bartosch 1997:67-68; Rösgen et al. 1997:9-12)
- the systematically lower efficiency of Otto engines compared to Diesel engines, in particular under partial load conditions.

For the latter reason, the increase of the fuel consumption for natural gas buses may vary depending on the route and the bus driver. However, even in the case of the six direct comparisons in Berlin (table 4.17), where both buses drove behind each other and the drivers switched buses halfway, the increase varies significantly. This indicates the importance of technical optimization of the natural gas buses. In particular, fuel consumptions might be reduced by up to 15 % by using lean burn motors instead of motors using a stoichiometric mix of gas and air (Kolke 1999:80). This corresponds to the lower value of 10 % increased fuel consumption relative to Diesel buses that will be used for sensitivity analyses. Care must be taken, however, that this lower fuel consumption is not achieved at the cost of significantly higher emissions of NO_x, which represent the dominant contribution to many impact categories, including human health (section 5.1).

The increased fuel consumption of the natural gas buses may in some cases even overcompensate the advantage of natural gas in terms of the specific CO_2 emission of 55 g CO_2 per MJ of net calorific value, compared to 74 g CO_2/MJ for Diesel. Taking into account the upstream emissions of climate gases as well as the methane emissions from the vehicles, which are both higher for natural gas, but small compared to the vehicle emissions of CO_2 in terms of CO_2 equivalents (see table 5.3), the break even point corresponds to an increased fuel consumption of natural gas buses of 23,3 % compared to Diesel buses, which is coincidentally almost equal to the average increased fuel consumption of 25 % found here.

Most studies comparing the emission factors of Diesel and natural gas buses are limited to the regulated pollutants PM, NO_x, VOC and CO. In particular, the large

Table 4.18. Fuel consumptions [MJ/km] and emission factors [g/km] for Diesel and natural gas buses (average inner-city driving pattern) used for the present case study (INFRAS 1999; Mangelsdorf et al. 1999:23; average of values in table 4.16 for CH_4 and NMVOC emissions of CNG buses, N_2O for CNG buses according to N_2O/CO_2 ratio for CNG in INFRAS 1998:102)

fuel / emission	mid 1980s[a]	EURO2[a]	Diesel EURO 3[a]	EURO 4[b]	natural gas
fuel consumption [MJ/km]	15,32	14,55	14,55	14,55	18,19
CO_2	1141	1084	1084	1084	1003
CH_4	0,052	0,026	0,021	0,018	1,7
N_2O	0,033	0,033	0,033	0,033	0,021
CO_2-equivalents	1152	1095	1094	1094	1045
PM 2,5	0,780	0,240	0,160	0,078	0,005
SO_2	0,216	0,205	0,205	0,027	0,000
NO_x (as NO_2)	18,0	10,8	7,2	3,6	2,2[d]
NMVOC	2,12	1,06	0,85	0,74	0,22
benzene	0,042	0,021	0,017	0,015	0,002
formaldehyde	0,130	0,063	0,051	0,044	0,003
acetaldehyde	0,047	0,023	0,019	0,016	0,001
B[a]P	7,2 E-6	3,6 E-6	2,9 E-6	2,5 E-6	2,5 E-6
1,3-butadiene	0,044	0,022	0,018	0,016	0,001

[a] reference year 2000 (sulfur content 300 ppb).
[b] reference year 2005 (sulfur content 40 ppb).
[c] Estimate in (Mangelsdorf et al. 1999:23) based on the CNG/Diesel emission ratios from (Rijkeboer and Hendriksen 1993), which appears to be more consistent with the values measured for vehicles (table 4.16) than their downward correction to 1,0 g NO_2/km based on a measurement for an engine.

group of volatile organic compounds is not further differentiated into substance groups or single substances with particular relevance for human health. Since a consistent set of emission factors for most relevant substances (as defined in section 4.1.3) for both Diesel and natural gas buses was derived by IFEU (Institut für Energie- und Umweltforschung, Heidelberg, Mangelsdorf et al. 1999), these will be used in the following, supplemented by emission factors for the remaining pollutants (CH_4, NMVOC) and fuel consumptions based on (INFRAS 1999) and the values listed in tables 4.16 and 4.17. These emission factors offer the further advantage of being differentiated according to the emission control standard for Diesel buses (1980s, EURO2 to EURO 4), such that the anticipated future development of the ‚conventional' technology (Diesel) can be taken into account when comparing it to the natural gas alternative[16].

In the case of Diesel buses, the IFEU emission factors for particulate matter, NO_x and VOC are largely identical to those from the standard reference (INFRAS

---

[16]The EURO 2 and 3 emission factors taken from (INFRAS 1999) refer to the year 2000, while those for the EURO 4 standard are based on the reference year 2005, which leads to a small additional emission reduction of some pollutants (PM, SO_2, NO_x) due to the assumed lower sulfur content of the Diesel fuel (30 ppm instead of 400 ppm, INFRAS 1999:57-8).

1999) for an average inner-city driving pattern, zero longitudinal inclination and the mix of standard buses (< 20 tons) and articulated buses (> 20 tons) with average passenger loads[17]. With regards to the emission factors for individual carcinogenic NMVOC, split factors were applied to the VOC emissions from (INFRAS 1999).

The IFEU emission factors for natural gas buses were derived by multiplying the emission ratios between CNG and Diesel determined from measurements for one passenger car and one light-duty vehicle (Rijkeboer and Hendriksen 1993) with the emission factors for the Diesel buses. These values were then adjusted to account for (unpublished) results of measurements of a natural gas bus motor carried out within a demonstration program of the German Federal Environmental Agency. In particular, the NO_x emission factor of 1,0 g/km was estimated by IFEU as a compromise between a value of 2,2 g/km according to the ratios in (Rijkeboer and Hendriksen 1993) and a value of 0,6 g/km according to a measurement of an *engine* for the German Federal Environmental Agency. Since the latter appears to be a particularly optimistic estimate compared to the values measured for buses in table 4.16, the higher value of 2,2 g/km will be used here instead. The emission factor of 1,0 g NO_x/km will be considered in a sensitivity analysis in chapter 5.

The emission factors for methane (CH_4) and NMVOC for the CNG buses were estimated according to the averages of the values in table 4.16. For the Diesel buses, they were taken from (INFRAS 1999). With regards to the fuel consumptions, the average percentage increase of 25 % (table 4.17) compared to the Diesel

Table 4.19. Cumulative Energy Demands (*CED*) [MJ/km], total emissions [g/km] and shares of the upstream processes [% of total] for Diesel buses (EURO 2 emission control standard) and natural gas buses (large fuel stations, fuel consumption 25 % higher than for the Diesel buses)

| | Diesel | | natural gas | |
	total	upstream [%]	total	upstream [%]
CED [MJ/km]	15,83	(8)	20,81	(9)
CO_2	1177	(8)	1158	(9)
CH_4	0,254	(90)	4,9	(65)
N_2O	0,035	(5)	0,027	(15)
CO_2-eq.	1194	(8)	1269	(14)
PM 10	0,017	(100)	0,028	(100)
PM 2,5	0,262	(8)	0,005	(0)
SO_2	0,541	(62)	0,110	(100)
NO_x	11,3	(3)	2,5	(10)
NMVOC	1,3	(19)	0,4	(38)
benzene	0,021	(0)	0,002	(0)
formaldehyde	0,063	(0)	0,003	(0)
acetaldehyde	0,023	(0)	0,001	(0)
B[a]P	3,6 E-6	(0)	5,0 E-8	(0)
1,3-butadiene	0,022	(0)	0,001	(0)

[17] In fact, IFEU was involved in the generation of the emission factors in (INFRAS 1999).

reference values were used rather than the averaged absolute consumptions, since these depend strongly on the driving conditions. This relative increase is in accordance with (INFRAS 1998:121). It was applied to the fuel consumption of Diesel buses from (INFRAS 1999). Due to the crucial importance of the fuel consumption, alternative increases of 10 % and 40 % will be used for sensitivity analyses. The CO_2 emissions were determined from the fuel consumptions by using the specific CO_2 emissions of 55 g/MJ for natural gas and 74 g/MJ for Diesel in Germany (Rausch et al. 1998). The SO_2 emission factor for Diesel buses was taken from (INFRAS 1999). It was set to zero for natural gas buses, since natural gas does not contain sulfur. The complete set of emission factors and fuel consumptions is shown in table 4.18.

4.3.2
Combination of Upstream and Vehicle Emissions

By multiplying the fuel consumptions from table 4.18 with the upstream emissions given in table 4.15, the latter can be related to a specific distance of driving. In this way, both upstream and vehicle emissions are expressed in mass per kilometer, such that they can be compared as well as added to yield the total emissions per kilometer (table 4.19 and fig. 4.4). In the case of the Diesel buses, they are based

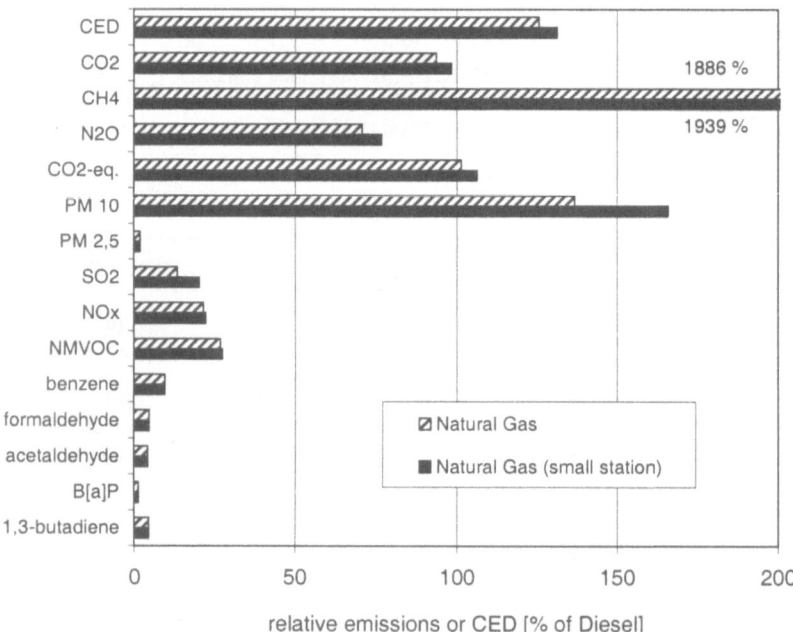

Fig. 4.4. Total emissions and Cumulative Energy Demand (*CED*) per driving distance for natural gas buses (large and small fuel station) relative to those of Diesel buses according to the data in table 4.19.

on the EURO 2 emission factors, in the case of natural gas buses on an increased fuel consumption of 25 % compared to Diesel buses and a large fueling station. The effect of variations of the emission control standard of Diesel buses, the fuel consumption of natural gas buses or the energy required for the compression of the natural gas at the fuel station will be considered in chapter 5 where necessary. For most pollutants, the vehicle emissions dominate the overall emissions. This holds true with the exception of CH_4 and SO_2 in both cases and of particulate matter in the case of natural gas buses.

A graphical comparison of the total emissions and energy demands per kilometer of driving for the natural gas buses relative to those for the Diesel buses is presented in fig. 4.4. The Cumulative Energy Demand and the climate gas emissions aggregated into CO_2-equivalents are higher for the natural gas buses, mainly because of their higher fuel consumption. The higher PM 10 emissions for natural gas are not significant, since they are by far overcompensated by the lower PM 2,5 emissions both in terms of the emitted mass as well as the associated health effects, as will be shown in chapter 5. The emissions of all remaining pollutants only amount to between 2 and 27 % of those for the Diesel buses. Since the upstream emissions typically only represent about 10 % or less of the total emissions, the difference between large fuel stations and small ones, which are also shown in fig. 4.4, are generally in the order of 5 %.

The upstream emissions of the non-methane volatile organic compounds (NMVOC) are not differentiated according to individual carcinogenic substances in (Rausch et al. 1998) as it is done here in the case of the vehicle emissions. Therefore, a direct comparison of the upstream and the vehicle emissions of the carcinogens is not possible.

The aggregated upstream NMVOC emissions are smaller than the vehicle emissions of NMVOC. Furthermore, most of the upstream emissions result from leakages of natural gas and from evaporative losses in the case of raw oil and are therefore likely to consist mostly of harmless substances such as alkanes. In the case of Diesel, a significant contribution also comes from the motors of ocean ships transporting the raw oil from OPEC countries to Germany. While the combustion exhausts of the ships may contain harmful substances (such as benzo[a]pyrene), the effective population density is low. For these reasons, the upstream emissions of the carcinogenic NMVOC considered here can be assumed to be negligible in a first approximation.

4.4
Cars

4.4.1
Vehicle Operation

Emission factors and fuel consumptions for cars fueled by Diesel, petrol and natural gas for all relevant air pollutants are available from (Bach et al. 1998) and are

Table 4.20. Fuel consumptions [MJ/km] and emission factors [g/km] for Diesel, petrol and natural gas cars (inner-city part of the European driving cycle) used for the present case study (Bach et al. 1998)

fuel / emission	Diesel	petrol	natural gas
fuel consumption [MJ/km]	3,19	3,32	3,65
CO_2[a]	238	248	201
CH_4	0,010	0,020	0,500
N_2O[b]	0,000	0,000	0,000
CO_2-equivalents	238	248	212
PM 2,5	0,050	9 E-5[c]	3 E-5[c]
SO_2	0,046	0,031	0,000
NO_x (as NO_2)	0,50	0,14	0,06
NMVOC	0,12	0,38	0,08
benzene	0,003	0,022	0,000
formaldehyde	0,020	0,002	0,0002
acetaldehyde	0,007	0,001	0,0001
B[a]P[d]	0,0 E+0	0,0 E+0	0,0 E+0
1,3-butadiene	0,002	0,003	0,000

[a] determined by multiplying the fuel consumptions with the CO_2 emission factors of the three fuels from (Rausch et al. 1998).

[b] set to zero because measured values were close to the sensitivity of the instruments for all three fuels.

[c] estimated from the ratios of the number concentration of particles in the exhausts measured for driving at a constant velocity of 50 km/h (Bach et al. 1998:40), which were multiplied with the measured emission factor of PM 2,5 for the Diesel cars.

[d] set to zero for a lack of measurement data, since contribution to overall health effects turns out to be negligible in the case of buses (chapter 5).

reproduced in table 4.20[18]. They are based on measurements on a test stand for the *same* mid-size car (a Renault Express) and refer to the inner city (ECE) part of the new European driving cycle according to EU Guideline 93/116 and an initial oil temperature of 25 °C. The increased mass of the natural gas car of 1300 kg compared to 1190 kg for the Diesel and petrol versions was thereby taken into account. While the use of test stand data may not exactly represent the absolute emissions under actual driving conditions, the parallel measurements for the same vehicle type provide a good basis for a comparative assessment.

As in the case of buses, the fuel consumption is higher for natural gas cars than for Diesel cars, in this case by 14 %, and also 10 % higher than for petrol cars. Since the increase of the fuel consumption is smaller than in the case of buses, however, the natural gas cars retains an emission advantage in terms of CO_2 and, despite the highest CH_4 emissions among the three fuels, also in terms of the CO_2-equivalents. The natural gas car also exhibits the lowest emissions of all other

[18] Emission factors for benzo[a]pyrene are not available. Since B[a]P emissions contribute only negligible amounts to the overall health effects of buses, they will be set to zero for cars.

pollutants. Between Diesel and petrol, each fuel is associated with higher emissions for some pollutants and lower emissions for others.

4.4.2
Combination of Upstream and Vehicle Emissions

Multiplication of the fuel consumptions of the cars (table 4.20) with the upstream emissions per lower calorific value of the fuel (table 4.15) and adding the vehicle emissions from table 4.20 yields the total emissions per kilometer of driving (table 4.21 and fig. 4.5). While the vehicle operation dominates the total emissions for many pollutants, the upstream emissions of methane, nitrous oxide and sulfur dioxide dominate for all three fuels. Furthermore, the emissions of particulate matter (PM 10 and PM 2,5 taken together) are dominated by the upstream processes for petrol and natural gas. For the petrol cars, the upstream NMVOC emissions appear to be larger than the vehicle emissions. However, since part of the vehicle emissions, namely the evaporative losses, were subsumed under the up-

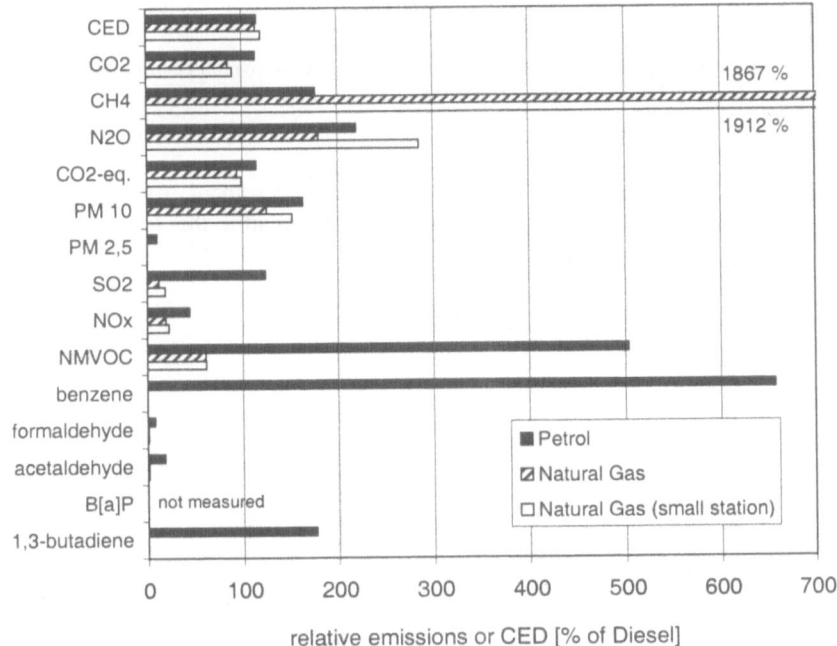

Fig. 4.5. Total emissions and Cumulative Energy Demands (*CED*) per driving distance for cars fueled by petrol and natural gas (large and small fuel station) relative to those for Diesel cars according to the data in table 4.21.

Table 4.21. Cumulative Energy Demands (*CED*), total emissions [g/km] and shares of the up-stream processes [% of total] for passenger cars fueled by Diesel, petrol and natural gas

	Diesel		petrol		natural gas	
	total	upstream [%]	total	upstream [%]	total	upstream [%]
CED [MJ/km]	3,47	(8)	4,03	(18)	4,18	(9)
CO_2	258	(8)	296	(16)	232	(9)
CH_4	0,060	(83)	0,107	(81)	1,149	(55)
N_2O	0,0004	(100)	0,001	(100)	0,001	(100)
CO_2-eq.	259	(8)	298	(17)	257	(14)
PM 10	0,004	(100)	0,006	(100)	0,006	(100)
PM 2,5	0,055	(9)	0,006	(98)	0,000	(0)
SO_2	0,120	(62)	0,148	(79)	0,022	(100)
NO_x	0,57	(12)	0,25	(44)	0,13	(45)
NMVOC	0,18	(32)	0,88	(57)	0,11	(25)
benzene	0,003	(0)	0,022	(0)	0,000	
formaldehyde	0,020	(0)	0,002	(0)	0,0002	(0)
acetaldehyde	0,007	(0)	0,001	(0)	0,0001	(0)
B[a]P[a]	0,0 E+0		0,0 E+0		0,0 E+0	
1,3-butadiene	0,002	(0)	0,003	(0)	0,000	

[a] set to zero for a lack of measurement data, since contribution to overall health effects turns out to be negligible in the case of buses (chapter 5).

stream emissions, the vehicle emissions are actually dominant. As in the case of the vehicle emissions alone, the natural gas cars exhibit the lowest emission factors among the three fuels, except for methane.

5 Life Cycle Assessment of Natural Gas Vehicles: Impact Assessment and Interpretation

5.1
Impact Assessment

In this section, the impacts associated with the energy conversion chains of natural gas, Diesel and petrol vehicles will be assessed for those impact categories selected in section 4.1, namely extraction of abiotic resources, climate change, acidification, nutrification, human toxicity and photo-oxidant formation. Most attention will be given to the site-dependent assessment of human health impacts, since this is where the method presented in chapter 3 is applied. For the other impact categories, generic assessments based on methods from the literature (section 2.5.2) will be provided. In the case of acidification, a more recent method including spatial differentiation will also be considered.

5.1.1
Site-dependent Assessment of Human Health Impacts

5.1.1.1
Upstream Processes

Among the pollutants considered for the upstream processes, effects on human health are associated with particulate matter (PM), secondary sulfate aerosols formed from emissions of SO_2, secondary nitrate aerosols formed from emissions of NO_x and ozone formed from NO_x and NMVOC as precursors. For particulate matter, sulfate aerosols and nitrate aerosols, spatially differentiated population exposures for the upstream processes were determined in chapter 4 (sections 4.2.1.3 and 4.2.2.3)[1]. With regards to particulate matter, a differentiation between emissions from Diesel motors leading to immissions of PM 2,5 and other emissions leading to immissions of PM 10 was thereby made due to the different effect factors (table 2.7). The health impacts of ozone formed from NO_x and NMVOC formation were only considered in a generic, site-independent way using the European average population exposures according to Hofstetter (1998) given in table 2.8. The human health impacts associated with the supply of 1 TeraJoule of

---

[1] Due to the necessity to consider a large number of upstream processes, it was more convenient to discuss the determination of the population exposures in connection with the inventory data, even though that is part of the impact assessment phase. The population exposures associated with the vehicle emissions, on the other hand, are determined in this chapter.

Table 5.1. Damages to human health from the upstream emissions of Diesel, petrol and natural gas per 1 TeraJoule of end energy (*EE*) provided at a fuel station in Germany

primary pollutant	secondary pol- lutant (or effect)	DALY / EE			YLD/ DALY [%]	YLL/ DALY [%]
		Diesel [a/TJ]	petrol [a/TJ]	natural gas [a/TJ]		
PM 10	(carcinogenic)	6,0 E-6	9,2 E-6	2,1 E-6	2	98
PM 10	(respiratory)	9,7 E-5	2,1 E-4	8,3 E-5	32	68
PM 2,5	(respiratory)	2,3 E-4	2,4 E-4	0,0 E+0	33	67
SO$_2$	sulfate aerosol	6,1 E-4	1,4 E-3	1,9 E-4	33	67
NO$_x$	nitrate aerosol	6,9 E-4	1,4 E-3	7,3 E-4	32	68
NO$_x$	ozone	2,9 E-5	4,4 E-5	1,8 E-5	66	34
VOC[a]	ozone	2,3 E-5	2,0 E-4	1,2 E-5	66	34
SUM		1,7 E-3	3,6 E-3	1,0 E-3	33-35[b]	65-67

DALY Disability Adjusted Life Years, *YLD* Years Lived Disabled, *YLL* Years of Life Lost, *POCP* Photochemical Ozone Creation Potentials
[a] using a POCP of 2,2 % for NMVOC emissions from the transportation and processing of crude oil and of 1,1 % for natural gas leaks (table 2.8).
[b] 33 % for natural gas, 34 % for Diesel, 35 % for petrol.

fuel (Diesel, petrol and natural gas) provided at a fuel station in Germany, which result from combining these population exposures with the effect factors from table 2.7, are shown in table 5.1 and fig. 5.1.

For each fuel, the health effects measured in terms of Disability Adjusted Life Years (DALY) are dominated by nitrate aerosols and sulfate aerosols, followed by primary particulates (PM 10 and PM 2,5), in the case of petrol along with ozone formed from NMVOC emissions. The same ranking also applies to the Years of Life Lost (YLL) and the Years Lived Disabled (YLD) separately. Due to the high energy demand and associated emissions of the refining process, which furthermore occur in Germany with its high population density, the supply of petrol is associated with human health impacts more than twice as high as those for Diesel and more than three times as high as those for natural gas supplied at large fuel stations. The effects for natural gas supplied at small stations are practically equal to those for the provision of Diesel. Furthermore, ozone formation only contributes significantly to the overall health effects in the case of petrol due to the high NMVOC emissions (table 4.15). In sections 5.1.1.3 and 5.1.1.4, the health impacts per fuel energy will be combined with the specific fuel consumptions to obtain the damage per distance of driving.

5.1.1.2
Damage Factors for Vehicle Emissions

The assessment of the health effects of the vehicle emissions uses the damage factors for the various pollutants in table 5.2 expressed in terms of Years Lived Disabled (YLD), Years of Life Lost (YLL) and Disability Adjusted Life Years (DALY) per mass M of emitted pollutant. These damage factors are obtained by

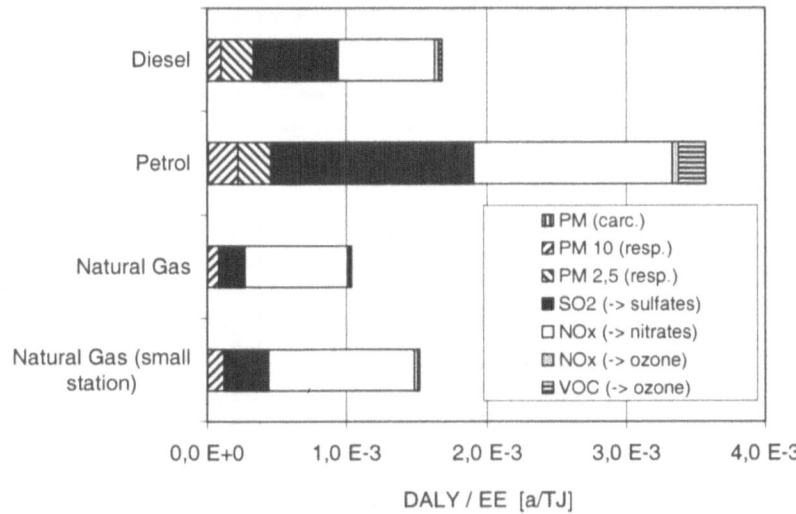

Fig. 5.1. Damages to human health in terms of Disability Adjusted Life Years (*DALY*) from the upstream emissions of Diesel, petrol and natural gas (large station and small station) per 1 Tera-Joule of end energy (*EE*) provided at a fuel station in Germany. The contributions of particulate matter (*PM*) via its carcinogenic effects and of NO$_x$ via ozone formation are too small to be visible.

multiplying the effect factors provided in tables 2.6 and 2.7 with the population exposures per mass of emitted pollutant calculated listed in table A.2 in the appendix.

For the primary pollutants, spatially differentiated damage factors are provided for large cities in agglomerations (settlement structure class (I,1)), „normal cities" in urbanized regions (II,1) and small municipalities in urbanized regions (II,3) within Germany according to the spatial specification of the functional unit in section 4.1.4. The ratio of the damage factors for large and small cities, which is indicated in the last column of table 5.2, ranges from 1,7 for the primary pollutant with the longest residence time (benzene, $\tau_a = 9$ days) to about 3,0 for short-lived pollutants (formaldehyde, acetaldehyde, $\tau_a = 6$-9 hours). For effects occurring through secondary sulfate and nitrate aerosols, the site dependence is relatively weak and is therefore not considered here. The desirable spatially differentiated consideration of the effects of ozone, which can be expected to vary by up to a factor of four in Europe (Spadaro and Rabl 1999:233), is a topic for further research. However, the contribution of ozone to the overall health effects will turn out to be negligible, such that a generic treatment is justified within this case study.

As in the case of the upstream emissions, the health effects of ozone were determined from the aggregated NMVOC emissions of the vehicles. This is because the individual organic compounds considered here were selected with regards to their carcinogenic properties. While individual POCP values can be assigned to them, this would only yield a more precise quantification of ozone formation if

Table 5.2. Spatially differentiated damage factors in terms of Disability Adjusted Life Years (*DALY*) per mass *M* of emitted substance for human health effects of traffic emissions of selected pollutants

primary pollutant	secondary pollutant (or effect)	YLD/ DALY [%]	YLL/ DALY [%]	DALY/M large city [a/kg]	DALY/M normal city [a/kg]	DALY/M small city [a/kg]	large/ small city
PM 2,5	(carcinogenic)	2	98	2,1 E-5	1,4 E-5	1,1 E-5	1,8
PM 2,5	(respiratory)	33	67	1,3 E-3	8,9 E-4	7,4 E-4	1,8
SO$_2$	sulfate	33	67		7,4 E-5		1,0
NO$_x$	nitrate	32	68		7,4 E-5		1,0
NO$_x$	ozone	66	34		1,3 E-6		1,0
NMVOC[a]	ozone	66	34		1,3 E-6		1,0
CH$_4$	ozone	66	34		1,3 E-8		1,0
benzene		3	97	4,3 E-6	2,9 E-6	2,5 E-6	1,7
formaldehyde[b]		3	97	4,2 E-6	2,1 E-6	1,4 E-6	3,0
acetaldehyde		4	96	7,9 E-7	4,0 E-7	2,7 E-7	2,9
benzo[a]pyrene[b]		2	98	4,7 E-2	2,8 E-2	2,2 E-2	2,2
1,3-butadiene		4	96	1,0 E-4	5,1 E-5	3,6 E-5	2,8

YLD Years Lived Disabled, *YLL* Years of Life Lost, *POCP* Photochemical Ozone Creation Potential

[a] based on the average POCP value of 59,2 % for NMVOC listed in table 2.8.

[b] For formaldehyde and benzo[a]pyrene, 63 % and 96 %, respectively, of the total damage is due to the uptake of food and drinking water (Hofstetter 1998:430).

numerous other NMVOC were considered in addition. In part due to a lack of the corresponding emission factors, in particular for the natural gas vehicles (in the case of buses[2]), and in part because the health effects of ozone turn out to be negligible, this was not done here. Estimates of the average POCP values applying to the vehicle emissions were not available either. The assessment of the ozone formation through the vehicle emissions was therefore based on the average POCP value of 59,2 % for NMVOC listed in table 2.8. The application of this country-average value (for the UK) to traffic emissions appears reasonable because traffic emissions typically represent a large share of the total NMVOC emissions of a country.

As can be seen from table 5.2, the damage through particulate matter is dominated by the respiratory health effects. This would also apply if the carcinogenic unit risk for Diesel soot particles was doubled according to (LAI 1992) compared to the value used here (see table 2.5). The dominant pathway for damage through nitrogen oxides is the formation of secondary nitrate aerosols. It should be borne in mind, however, that the epidemiological evidence for the health effects of secondary nitrate aerosols is weaker than that for primary particles and for ozone.

---

[2] Emission factors for individual NMVOC contributing to ozone formation are provided in (Bach et al. 1998:18-25) for cars fueled by Diesel, petrol and natural gas.

5.1.1.3
Buses

Category indicator results for human health are provided for the reference cases of Diesel buses with EURO 2 emission control standard and natural gas buses with a fuel consumption increased by 25 % compared to the Diesel bus and fueled by a large fuel station. The respective emissions are listed in table 4.18. The sensitivity of the overall results to variations of the input parameters will be considered in section 5.2.

The damage factors in table 5.2 can be multiplied with the emission factors for the buses from table 4.18 to obtain the damages to human health per kilometer of driving associated with the various pollutants emitted by the vehicles. The contribution of the upstream emissions is obtained by multiplying the damages per fuel energy (table 5.1) with the fuel consumptions of the vehicles (table 4.18). Since the contribution of the upstream emissions to the overall health effects will turn out to be in the order of 10 % or less, they will not be differentiated according to individual pollutants in the following, keeping in mind that the dominant upstream contributions come from nitrate and sulfate aerosols. The contributions of the various pollutants emitted by the vehicles and the aggregated contribution of the upstream emissions to the human health effects, expressed in terms of DALY per distance of driving, are shown in figures 5.2 and 5.3. In both figures, emissions in large cities (settlement structure class (I,1)) and small cities (class (II,3)) are distinguished in order to examine the spatial differentiation of the impacts.

In fig. 5.2, the contributions of the various pollutants emitted by the vehicles and the upstream emissions are plotted on a logarithmic scale, since they vary over six orders of magnitude. For both Diesel buses and natural gas buses, the dominant contribution to the overall health impacts comes from nitrate aerosols. The second rank is assumed by Diesel soot particles in the case of the Diesel buses and by the upstream emissions (the health effects of which are themselves dominated by nitrate aerosols) in the case of the natural gas buses. In both cases, small contributions come from ozone and sulfate aerosols (upstream only for natural gas), and the effect of the carcinogenic substances can be neglected[3]. In particular, it is interesting to note that the upstream emissions of some pollutants (primary and secondary particles) yield larger contributions to the overall health effects than the vehicle emissions of the carcinogens. This shows that the upstream emissions cannot, in general, be neglected. The disregard of the upstream emissions of the considered carcinogenic substances is nevertheless justified by the fact that they are small compared to the vehicle emissions which themselves contribute only insignificantly to the overall health effects.

Fig. 5.3 shows a comparison of the overall health impacts of the Diesel and natural gas buses, expressed in terms of Disability Adjusted Life Years (DALY). The same relations as for the DALY indicator also apply to the indicators Years of Life Lost (YLL) and Years Lived Disabled (YLD) separately (see table 5.3). This

---

[3] For formaldehyde and benzo[a]pyrene, this holds even though 63 % and 96 %, respectively, of the total damage is due to the uptake of food and drinking water (Hofstetter 1998:430), which is not shown in fig. 5.2.

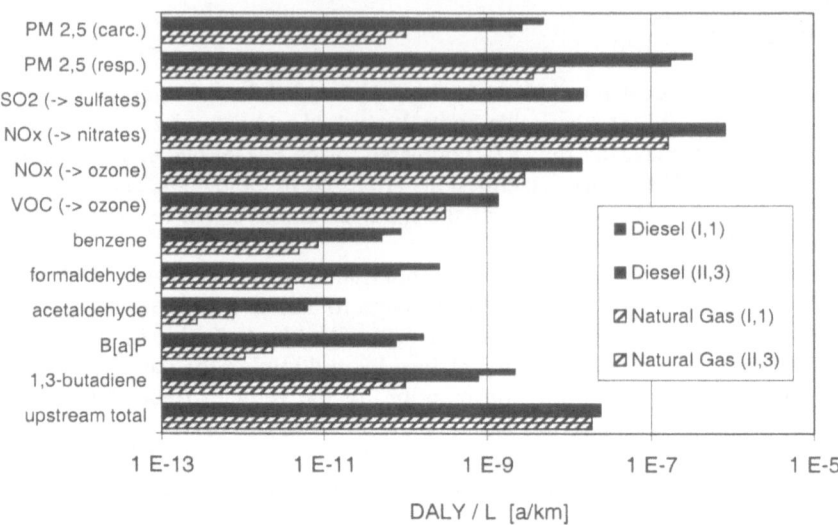

Fig. 5.2. Contributions of individual pollutants emitted by the vehicles and of the aggregated upstream emissions to the human health impacts for Diesel buses and natural gas buses, expressed as Disability Adjusted Life Years (*DALY*) per distance *L* of driving and differentiated according to the location of driving: *I,1* large cities in agglomerations, *II,3* small cities in urbanized regions

is because the YLL/DALY and YLD/DALY ratios are the same for the dominating contributions from primary particles and secondary nitrate aerosols.

The health impacts of the natural gas buses are independent of the size of the city, since they do not contain any significant contributions from primary pollutants. Even the impacts of the Diesel buses depend only weakly on the location of driving, since the contribution of the secondary nitrates is higher than that of the primary Diesel particles. The health impacts of the natural gas buses are lower by a factor 6,3 in large cities and by a factor 5,5 in small cities than those of the Diesel buses. Therefore, the substitution of Diesel by natural gas is almost as worthwhile in small cities as it is in large cities within agglomerations.

An important qualification has to be made concerning the nitrate aerosols. If nitrate aerosols can indeed be associated with the same health effects as primary particles, as it is assumed here, these effects dominate the overall health effects of both Diesel and natural gas buses. If it should turn out, on the other hand, that the association between nitrate aerosols and health effects cannot be maintained (in which case it is likely to be either weaker or non-existing), the overall impacts of both Diesel and natural gas buses are likely to be dominated by the respiratory health effects of primary fine particulates. In that case, the health impacts of the natural gas buses would be lower by a factor 24 in large cities and by a factor 18 in small municipalities than those of the Diesel buses. In particular, there would be a clear difference between large cities and small municipalities.

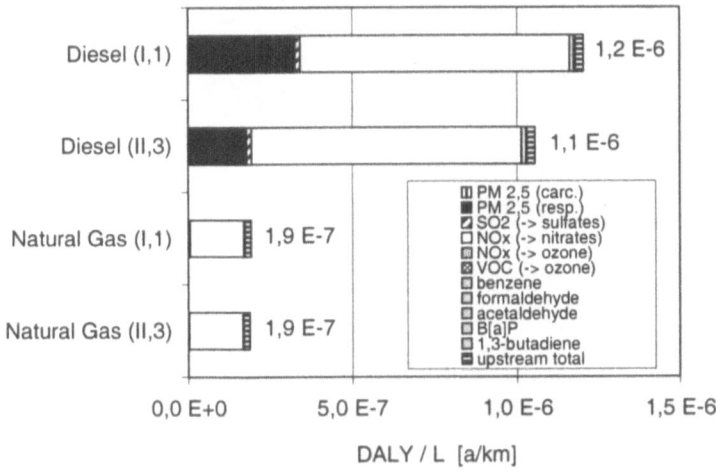

Fig. 5.3. Total human health impacts for Diesel buses and natural gas buses, expressed as Disability Adjusted Life Years (*DALY*) per distance *L* of driving and differentiated according to the location of driving: *I,1* large cities in agglomerations, *II,3* small cities in urbanized regions. The contributions of most pollutants are too small to be visible (see fig. 5.2).

5.1.1.4
Cars

Fig. 5.4 shows the contributions of the individual pollutants emitted by the cars and of the aggregated upstream emissions to the overall health impacts. For reasons of graphical clarity, only the results for the case of large cities in agglomerations are shown. Compared to the example of buses (fig. 5.2), the most important difference is that the respiratory health effects of primary fine particles (PM 2,5) are now the dominating contribution to the overall health effects of the Diesel vehicles. According to the damage factors in table 5.2, the health effects of secondary nitrate aerosols and primary fine particles would be equal for a ratio of the emission factors NO_x/PM 2,5 of 17 in the case of large cities (I,1), of 12 for ‚normal cities‘ (II,1) and of 9 for small municipalities (II,3). For the Diesel cars considered here, the emission ratio NO_x/PM 2,5 amounts to 10 (table 4.20), such that the respiratory health effects of primary fine particles dominate for large and normal cities, while the impacts of nitrates are somewhat bigger for small municipalities. The Diesel buses considered above, on the other hand, are characterized by a high NO_x/PM 2,5 emission ratio of 45 (table 4.18), such that the health effects of the secondary nitrates dominate irrespective of the location of driving. For the petrol and natural gas cars, it should furthermore be noted that the contribution of the upstream emissions is almost as high as that of the vehicle emissions taken together.

The total health impacts aggregated over all contributing pollutants are largest for the Diesel cars as a result of the highest emission· factors for both primary fine

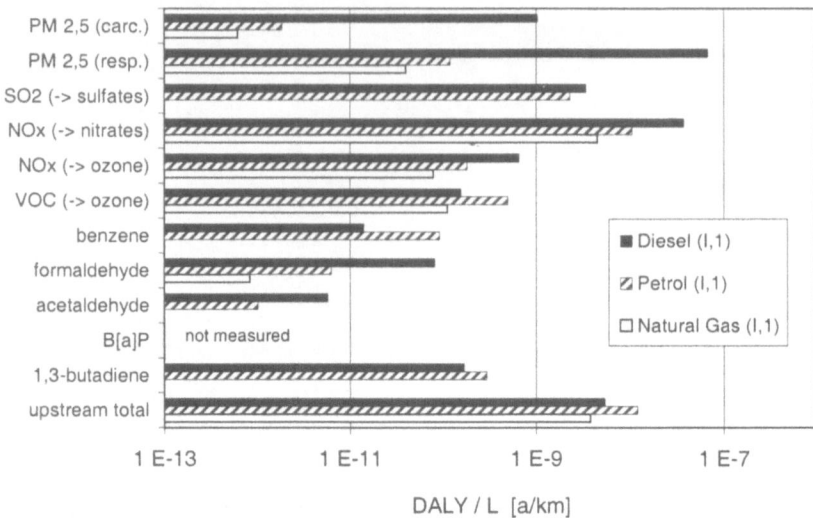

Fig. 5.4. Contributions of individual pollutants emitted by the vehicles and of the aggregated upstream emissions to the human health impacts for Diesel, petrol and natural gas cars, expressed as Disability Adjusted Life Years (*DALY*) per distance L of driving for an operation of the vehicles in large cities in agglomerations (*l,1*). For formaldehyde, 63 % of the total damage is due to the uptake of food and drinking water (Hofstetter 1998:430), while only the impacts due to inhalation are shown here.

particulates (PM 2,5) and nitrogen oxides (NO_x) (fig. 5.5)[4]. As a result of the dominance of primary fine particulates for large and normal cities, the spatial differentiation of the health impacts of the Diesel cars of a factor 1,4 between large and small cities is larger than in the case of buses, where it amounts to a factor 1,1. The impacts of natural gas and petrol cars are independent of the size of the city. The total health impacts of the natural gas cars are smaller by a factor of 10 in small cities and by a factor of 14 in large cities compared to their Diesel counterparts. Accordingly, the substitution of Diesel cars by natural gas is more worthwhile in large cities within agglomerations than in small cities.

The total health impacts of the petrol cars are about three times lower than those for the Diesel cars due to lower emissions of both NO_x and fine particles. On the other hand, they are still three times as high as those for cars fueled by natural gas. This is due to the higher NO_x emissions in combination with the higher health impacts of the upstream processes. As in the case of buses, the lower quality of the evidence for the health effects of the secondary nitrate aerosols needs to be kept in mind.

---

[4] This and the following statements also apply to the indicators Years of Life Lost (YLL) and Years Lived Disabled (YLD) separately, as can be seen in table 5.6.

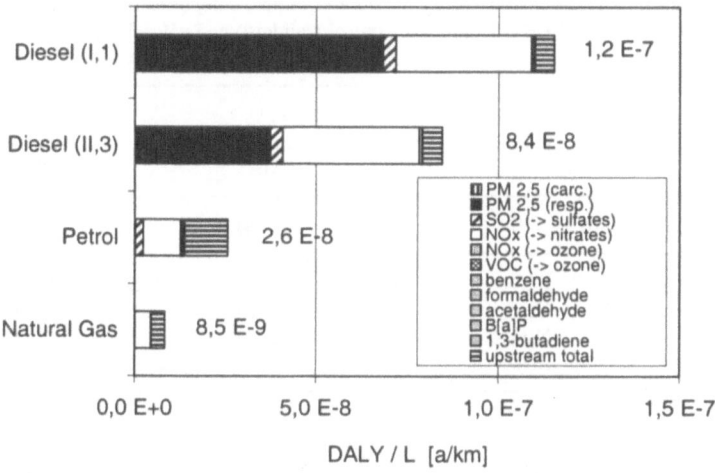

Fig. 5.5. Total human health impacts for Diesel, petrol and natural gas cars, expressed as Disability Adjusted Life Years (*DALY*) per distance *L* of driving and differentiated according to the location of driving in the case of Diesel: *I,1* large cities in agglomerations, *II,3* small cities in urbanized regions. The contributions of most pollutants are too small to be visible (see fig. 5.4).

5.1.2
Generic Assessment for Other Impact Categories

5.1.2.1
Buses

For the impact categories other than human toxicity, a standard impact assessment without spatial differentiation (except for acidification) was carried out by combining the total emission values from table 4.19 with the equivalency factors provided in section 2.5.2. The results are presented in table 5.3. In order to provide an overview of all considered impact categories, the human health impacts aggregated in terms of DALYs (section 5.1.1.3) are shown here again. It should be noted that they include the human health effects of ozone formation. In order to avoid double counting, the ethylene equivalents shown in table 5.3 should therefore only be used for other effects of ozone formation, e.g. on vegetation or agricultural crops.

In order to facilitate a qualitative comparison between the incommensurable indicators for the different impact categories, they were normalized to the corresponding reference values associated with one year of life of an average citizen of the European Union (table 5.4). The normalized category indicators are shown in fig. 5.6. Their values are in the order of 1 E-4 person-years/kilometer for the Diesel buses, which means that driving the bus for about 10,000 kilometers is associ-

Table 5.3. Category indicator results for all considered impact categories for the energy conversion chains of Diesel buses (EURO 2 emission control standard) and natural gas buses (large fuel stations)

impact category	indicator	pollutant	Diesel total	Diesel upstream [%]	natural gas total	natural gas upstream [%]
resources	CED [MJ/km]		15,83	(8)	19,90	(9)
climate change	CO_2-eq. [g/km]	CO_2	1177	(8)	1102	(9)
		CH_4	5	(90)	101	(65)
		N_2O	11	(5)	8	(15)
		SUM	1194	(8)	1210	(14)
acidification	SO_2-eq. [g/km]	SO_2	0,5	(62)	0,1	(100)
		NO_x	7,9	(3)	1,7	(10)
		SUM	8,5	(7)	1,8	(14)
nutrification	N-eq. [g/km]	NO_x	3,4	(3)	0,7	(10)
ozone formation	ethylene-eq. [g/km]	NO_x	6,70	(3)	1,45	(10)
		NMVOC	0,63[a]	(1)[b]	0,13	(1)[c]
		CH_4	0,00	(90)	0,03	(65)
		SUM	7,34	(3)	1,61	(10)
human toxicity	YLD [a/km]	SUM	3,5-4,0 E-7[d]	(2)	6,3 E-8	(10)
	YLL [a/km]	SUM	7,1-8,1 E-7[d]	(2)	1,3 E-7	(10)
	DALY [a/km]	SUM	1,1-1,2 E-6[d]	(2)	1,9 E-7	(10)

CED Cumulative Energy Demand, *YLD* Years Lived Disabled, *YLL* Years of Life Lost, *DALY* Disability Adjusted Life Years, *POCP* Photochemical Ozone Creation Potential
[a] average POCP for NMVOC according to table 2.8.
[b] POCP = 2,2 % for processing and transport of crude oil according to table 2.8.
[c] POCP = 1,1 % for leakages of natural gas according to table 2.8.
[d] range between small cities in urbanized regions and large cities in agglomerations.

ated with impacts approximately as high as those resulting from all activities of one average European person during one year[5].

With the exception of climate change and the extraction abiotic resources, the category indicators for the natural gas buses are between a factor 5 and 6 smaller than those for the Diesel buses, which is mainly due to the strong reduction of the respective vehicle emissions, in particular of NO_x, which represents the largest contributions to all of these impact categories. Based on the assumed increased

[5] To make the comparison meaningful, it should be noted that the driving distance of 10,000 vehicle-kilometers equals about 120,000 person-kilometers for an average occupancy rate of 12 persons per bus (Rausch et al. 1998), i.e. the annual environmental impact of one average EU-citizen corresponds to riding on a Diesel bus for about 120,000 kilometers.

Table 5.4. Reference values used for normalization (Hauschild and Wenzel 1998)

impact category	indicator	reference value	unit	comments
resources	CED	1,8 E+5	MJ / pers. a	primary energy demand in Germany (VDI 1998:538)
climate change	CO_2-eq.	1,9 E+4	kg / pers. a	for Denmark
acidification	SO_2-eq.	83	kg / pers. a	for European Union
nutrification	N-eq.	58	kg / pers. a	for Denmark
ozone formation	ethylene-eq.	51	kg / pers. a	average POCP of 0,4 for NMVOC instead of 0,592 used here (table 2.8)
human toxicity	DALY	8,7 E-3	a / pers. a	for Europe (ca. 700 Mio. inhabitants, Hofstetter 1998:342)

CED Cumulative Energy Demand, *pers.* persons, *eq.* equivalents, *POCP* Photochemical Ozone Creation Potential, *DALY* Disability Adjusted Life Years
[a] including emissions of 52,4 kg NO_x per capita and year (Denmark), multiplied with the POCP of NMVOC (59,2 %) due to the equality of the fate factors in table 2.8.

fuel consumption of 25 % for natural gas buses compared to Diesel buses, the CO_2-equivalents for both fuel chains are practically equal. More precisely, the break-even occurs for an increase of 23,3 % in the fuel consumption. Higher fuel consumptions would lead to disadvantages of the natural gas buses in terms of climate change, and vice versa.

The extraction of abiotic resources is the only impact category with the disadvantage being on the side of natural gas. This is mainly due to the increased fuel consumption in combination with the higher Cumulative Energy Demand per end energy of the provided fuel. Because of the latter, this disadvantage for natural gas would also remain, even though to a lesser extent, in the unrealistically optimistic case of a fuel consumption equal to that of the Diesel buses.

Due to the trade-off between the extraction of abiotic resources and all other impact categories in the comparison of natural gas and Diesel buses, a qualitative or quantitative determination of the relative importance of the various categories would in principle have to be made to arrive at an overall preference for either one of the fuels. However, since natural gas buses only hold a disadvantage in one impact category, and since the normalized impacts are all in the same order of magnitude, an overall preference for Diesel buses is quite unlikely. It would require an unrealistically larger weight being put on abiotic resource extraction compared to all the other impact categories, including human health[6]. To what

---

[6] The weight for the category of resources would need to be about 15 times the sum of the weights for the categories of acidification, nutrification, ozone formation and human toxicity. This holds exactly, if the same weight was assumed for all four impact categories, and approximately otherwise. The weights thereby refer to the reference situations used for the normalization, i.e. for an overall disadvantage for natural gas to occur, the current total European resource consumption would need to be considered at least 15 times as severe as the total European impacts due to acidification, nutrification, ozone formation and human toxicity taken together.

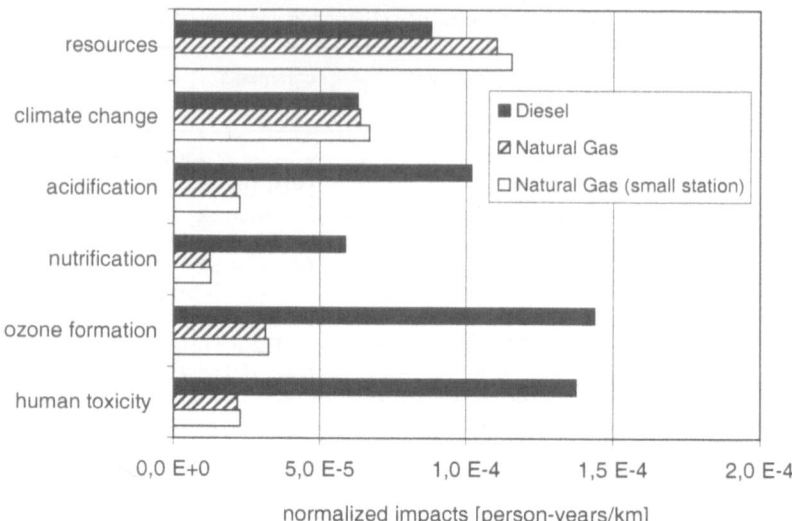

Fig. 5.6. Normalized impact indicators for the energy conversion chains of Diesel buses and natural gas buses for selected impact categories. The human toxicity indicators refer to driving the buses in large cities within agglomerations.

extent this would change if natural gas buses also held a disadvantage in terms of climate change will be discussed in section 5.2.

In addition to the above generic impact assessment, a spatially differentiated impact assessment was carried out for the impact category of acidification using the characterization factors of Potting et al. (1998) for the year 1990 (table 2.3). With regards to the upstream emissions, the information about the location of the emission processes contained in tables 4.6, 4.7, 4.11 and 4.12 was used. Processes occurring at sea were characterized by the arithmetic mean of the acidification factors for the Atlantic Ocean, the Mediterranean and the North Sea. The processes contributing less than 1 % to the overall upstream emissions, the locations of which were not identified, were associated with the default acidification factors from table 2.3. The vehicle emissions were assumed to occur in the western part of Germany. Table 5.5 lists the aggregated acidification factors for the upstream, vehicle and total emissions and the category indicator values (change in unprotected ecosystem area) resulting from their multiplication with the emissions of SO_2 and NO_x listed in table 4.18.

The differences between the acidification factors for individual processes of the two fuel supply chains of up to a factor 10 for SO_2 and up to a factor 3 for NO_x partially cancel out in the sum of the upstream processes, with differences of 18 % and 28 % remaining, respectively. The overall result, however, is dominated by the vehicle emissions of NO_x, to which the same acidification factors apply for both Diesel and natural gas vehicles. For this coincidental reason, the ratio of the overall category indicators for the two vehicle types (columns 3 and 6, rows 3 and 8) is nearly the same according to both the generic and the spatially differentiated

Table 5.5. Comparison of the generic impact assessment of acidification shown in table 5.3 and an impact assessment based on the spatially differentiated characterization factors for the year 1990 from (Potting et al. 1998) listed in table 2.3

		pollu-tant	Diesel			natural gas		
			upstream	vehicle	total	upstream	vehicle	total
generic	SO$_2$-eq. [g/km]	SO$_2$	0,3	0,2	0,5	0,1	0,0	0,1
		NO$_x$	0,2	7,7	7,9	0,2	0,7	0,9
		SUM	0,6	7,9	8,5	0,2	0,7	0,9
spatial dif-ferentiation	acid. factor [ha/ton]	SO$_2$	2,99	1,94	2,59	3,53	1,94	3,53
		NO$_x$	0,95	1,42	1,41	1,29	1,42	1,39
	unprotected area [1E-2 m^2/km]	SO$_2$	1,0	0,4	1,4	0,3	0,0	0,3
		NO$_x$	0,3	15,6	15,9	0,3	1,4	1,7
		SUM	1,3	16,0	17,3	0,6	1,4	2,0

eq. equivalents, *acid.* acidification

assessment, i.e. the spatially differentiated assessment coincidentally confirms the results of the generic assessment. The former is nevertheless useful even in this case since it provides more meaningful category indicators. The use of the acidification factors for the year 2010 leads to similar results, which are not shown here in detail.

5.1.2.2
Cars

In the same way as for the buses, a generic impact assessment for the impact categories other than human health was carried out for the cars fueled by Diesel, petrol and natural gas. The resulting category indicator values are listed in table 5.6 in absolute terms and are plotted as normalized values in fig. 5.7. Compared to petrol, natural gas fares equal or better with regards to all impact categories. Compared to Diesel, a disadvantage occurs only with regards to the extraction of abiotic resources, as in the case of buses. For natural gas cars to be considered less favorable than cars running on Diesel, the impact category of abiotic resources would need to be given a weight 13 times as high as the weights of the five remaining impact categories taken together, which is highly unplausible, given that the latter include both climate change and human toxicity. Therefore, natural gas cars are preferable compared to both Diesel and petrol cars with regards to the considered impact categories.

With regards to the comparison between Diesel and petrol, the latter holds disadvantages with regards to the extraction of resources and climate change mainly due to the higher energy demand for the refining. These need to be compared with the advantages in terms of acidification, nutrification and ozone formation due to the lower NO$_x$ emissions and the advantages in terms of human toxicity due to the lower emissions of both NO$_x$ and fine particulates. The determination of an overall

Table 5.6. Category indicator results for all considered impact categories for the energy conversion chains of cars using Diesel, petrol and natural gas

impact category	indicator	pollutant	Diesel total	Diesel upstream [%]	petrol total	petrol upstream [%]	natural gas total	natural gas upstream [%]
resources	CED [MJ/km]		3,47	(8)	4,03	(18)	3,99	(9)
climate change	CO₂-eq. [g/km]	CO₂	258	(8)	296	(16)	221	(9)
		CH₄	1	(83)	2	(81)	24	(55)
		N₂O	0	(100)	0	(100)	0	(100)
		SUM	259	(8)	298	(17)	245	(14)
acidification	SO₂-eq. [g/km]	SO₂	0,1	(62)	0,1	(79)	0,01	(100)
		NOₓ	0,4	(12)	0,2	(44)	0,1	(45)
		SUM	0,5	(24)	0,3	(60)	0,1	(54)
nutrification	N-eq. [g/km]	NOₓ	0,2	(12)	0,1	(44)	0,03	(45)
ozone formation	ethylene-eq. [g/km]	NOₓ	0,34	(12)	0,15	(44)	0,07	(45)
		NMVOCª	0,07	(2)	0,24	(5)	0,05	(1)
		CH₄	0,00	(83)	0,00	(81)	0,01	(55)
		SUM	0,41	(11)	0,38	(20)	0,12	(28)
human toxicity	YLD [a/km]	SUM	2,8-3,8 E-8ᵇ	(7-5ᵇ)	8,8 E-9	(47)	2,8 E-9	(44)
	YLL [a/km]	SUM	5,7-7,7 E-8ᵇ	(6-5ᵇ)	1,7 E-8	(45)	5,6 E-9	(45)
	DALY [a/km]	SUM	0,8-1,2 E-7ᵇ	(6-5ᵇ)	2,6 E-8	(46)	8,5 E-9	(45)

CED Cumulative Energy Demand, *YLD* Years Lived Disabled, *YLL* Years of Life Lost, *DALY* Disability Adjusted Life Years, *POCP* Photochemical Ozone Creation Potential
[a] POCP = 2,2 % for processing and transport of crude oil and 1,1 % for leakages of natural gas, average POCP of 59,2 % for NMVOCs emitted from vehicles according to table 2.8.
[b] range between small cities in urbanized regions and large cities in agglomerations.

preference is not trivial in this case and requires an ordering or weighting of the impact categories. This was not done, however, since the comparison between Diesel and petrol is not of the foremost interest here. Furthermore, since the site-dependent assessment of acidification did not lead to results very different from those of the generic assessment for the buses, it was omitted for the case of cars.

Finally, it should be noted that the contributions of the upstream processes to the overall impacts (table 5.6) are generally higher than in the case of buses. This is because the ratio of upstream and vehicle emissions of NOₓ is about five times as high for the cars than for the buses, with NOₓ being the pollutant of highest

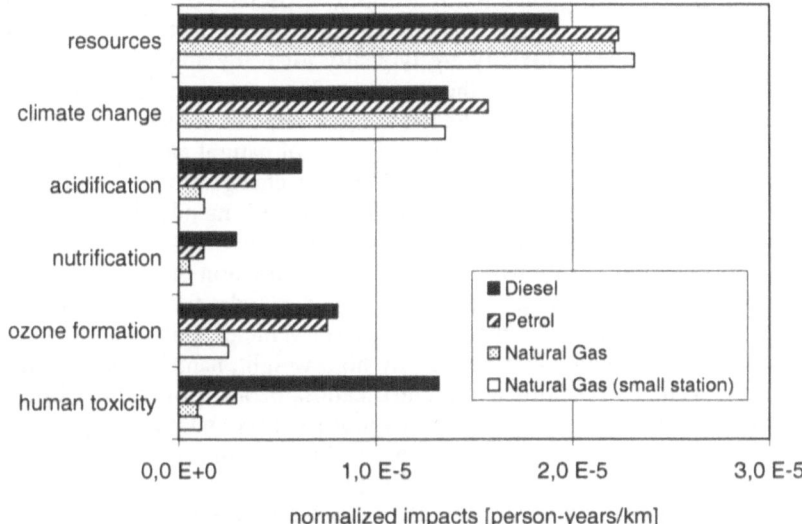

Fig. 5.7. Normalized impact indicators for the energy conversion chains of Diesel, petrol and natural gas cars for selected impact categories. The human toxicity indicators refer to driving in large cities within agglomerations.

importance for many impact categories (acidification, nutrification, ozone formation, human health).

5.2
Interpretation

According to ISO/DIS 14043, the interpretation phase of LCA includes the identification of essential contributions to the results of the inventory analysis and the impact assessment and an evaluation of these results in terms of completeness and sensitivity to particular input parameters (section 2.6). Many elements of the interpretation have been integrated into the goal and scope definition and the reporting of the results of the inventory analysis and impact assessment: The completeness of the results was discussed in sections 4.1.2 and 4.1.3 in relation to the definition of the system boundaries and the selection of impact categories and interventions to be considered. The discussion of significant contributions was integrated into the reporting of the inventory results in chapter 4 (sections 4.2.1.2, 4.2.2.2, 4.2.3, 4.3.2 and 4.4.2) and of the impact indicator results in this chapter. Therefore, after summarizing the overall outcome of the comparison of the environmental effects of the considered Diesel, petrol and natural gas vehicles, only the main factors influencing this outcome shall be recalled here, and corresponding sensitivity analyses carried out where necessary.

Main results: Due to strong emission reductions, especially of NO_x and PM, the impacts of the natural gas vehicles in the categories acidification, nutrification, ozone formation and human toxicity are typically lower by about a factor 5 than those of the Diesel vehicles, while the impacts of the petrol cars are inbetween. The increased fuel consumptions of the natural gas vehicles are such that they just about compensate the lower specific CO_2 emissions of natural gas, i.e. the impacts of natural gas and Diesel vehicles in terms of climate change are practically equal. The climate impacts of petrol cars are somewhat higher, mainly due to the high energy demand of the refining process for petrol. The only disadvantage of natural gas compared to Diesel occurs with regards to the extraction of abiotic resources as a result of the higher fuel consumption. However, in order for this disadvantage to shift the overall preference to the side of Diesel vehicles, this impact category would need to be attributed an unrealistically high weight, namely about 15 times the sum of the weights of acidification, nutrification, ozone formation and human toxicity. The overall comparison between natural gas cars and petrol cars is more straightforward, since natural gas fares equal or better with regards to all impact categories. Therefore, it can be concluded for both buses and cars that natural gas is the most favorable fuel with regards to the considered impact categories.

Influence of the fuel consumption: If the increase of the fuel consumption of the natural gas vehicles relative to the Diesel vehicles was higher than in the considered reference cases, e.g. 40 % instead of 25 % in the case of buses (table 4.17), a disadvantage of natural gas in terms of climate change would result in addition to the higher extraction of abiotic resources. For an overall disadvantage of natural gas to occur, the sum of the weights for the impact categories of resources and climate change would have to be about 9 times the sum of the weights for acidification, nutrification, ozone formation and human toxicity. This may still appear unplausible, but less so than in the case where no disadvantage for natural gas in terms of climate change exists.

Dominant contributions to human health impacts: The total impacts of Diesel vehicles on human health are dominated by the respiratory health effects of either the primary Diesel soot particles (PM 2,5) or by secondary nitrate aerosols formed from emissions of nitrogen oxides (NO_x). It is thereby supposed that the epidemiological evidence for the respiratory health effects of the nitrates is considered to be sufficient despite being weaker than in the case of the primary particles. While the effects of the Diesel soot particles vary by about a factor 1,7 between large cities and small municipalities in Germany, the effects of the nitrate aerosols vary on a larger geographical scale, but can be approximated by a constant value for all sites in Germany. Within this approximation, the effects of the two pollutants would be equal for a ratio of the emission factors NO_x/PM 2,5 of 17 in the case of large cities, 12 for ‚normal cities' and 9 for small municipalities.

The Diesel buses considered here are characterized by a high NO_x/PM 2,5 emission ratio of 45 (table 4.18), such that the health effects of the secondary nitrates dominate irrespective of the location of driving. This conclusion is likely to remain stable if the site-specific variations of the effects of the primary particles within the generic spatial classes (large, normal and small cities) and of the effects of the nitrates across Germany are taken into account, which are both characterized by standard deviations in the order of 45 % (sections 3.6.3 and 3.6.5). For the Diesel cars considered here, the emission ratio $NO_x/PM\,2,5$ amounts to 10 (table 4.20), such that the respiratory health effects of primary fine particles dominate for large and normal cities, while the impacts of nitrates are somewhat bigger for small municipalities. Due to the proximity of the $NO_x/PM\,2,5$ emission ratio to the break even points, this conclusion may change, however, if the abovementioned site-specific considerations are taken into account. For the natural gas buses and cars and the petrol cars, the overall health effects are dominated by the nitrate aerosols, such that they are independent of the size of the city.

Spatial differentiation of the health benefits of using natural gas: The health impacts of the natural gas buses are lower by a factor 6,3 in large cities and by a factor 5,5 in small cities than those of the Diesel buses. Therefore, the substitution of Diesel by natural gas is almost as worthwhile in small cities as it is in large cities within agglomerations. This is due to the dominant contribution of the nitrate aerosols to the overall health effects in both cases. In the case of cars, the total health impacts for natural gas cars are smaller by a factor of 10 in small cities and by a factor of 14 in large cities compared to Diesel. Accordingly, the substitution of Diesel cars by natural gas is more worthwhile in large cities in agglomerations than in small cities. This is due to the predominant contribution of the primary particles to the overall health effects of the Diesel buses. Nevertheless, the health benefits of the substitution of Diesel by natural gas are in the same order of magnitude for all types of cities. The health benefits of substituting petrol cars by natural gas cars are independent of the size of the city, since primary pollutants do not play a significant role in either case. These statements about the dependence of the health benefits of substituting Diesel or petrol by natural gas on the size of the city refer to the average impacts of the primary particles for the generic settlement structure classes and the average impacts of the secondary nitrate aerosols for Germany calculated in chapter 3. Since both averages are characterized by standard deviations in the order of 45 %, site-specific considerations are required in order for statements about individual emission sites to be made.

Influence of the emissions of nitrogen oxides (NO_x): The emissions of NO_x not only play an important role with regards to the human health effects, but also generally dominate the indicator values for the categories of acidification, nutrification and ozone formation. Besides the fuel consumptions, which largely determine the results for the categories of climate change and resources, the NO_x emission factors of the vehicles are therefore a crucial input parameter. In this context, it may be recalled that an emission factor of 2,2 g NO_2/km instead of 1,0 g NO_2/km was used for the natural gas buses (table 4.18). The use of the lower value would imply a reduction of the impacts of the natural gas buses in the categories of acidification, nutrification, ozone formation and human toxicity roughly in the order of 50 % (see table 5.7). Even if the epidemiological evidence for the respiratory health effects of secondary nitrate aerosols turned out to be invalid, the reduction of NO_x emissions would continue to be a worthwhile undertaking. In particular, if lean burn motors will be used to reduce the fuel consumption of natural gas buses, care must be taken that this is not achieved at the cost of significantly higher NO_x emissions. This holds in particular if future possible reductions of the NO_x emissions of Diesel buses according to the EURO 3 and EURO 4 emission control standards are taken into account (see below).

Significance and spatial differentiation of the upstream processes: The contributions of the upstream processes to the total values of the impact indicators are in the order of 10 % or less for buses running on either Diesel or natural gas (table 5.3). In the case of petrol cars, the upstream contributions to resource extraction and climate change are closer to 20 % due to the high energy demand of the refining process. Furthermore, the upstream contributions to the impacts of the petrol and natural gas cars in the categories of acidification, nutrification, ozone formation and human toxicity are mostly in the order of more than 40 % (table 5.6). For the most part, this is due to the higher share of the upstream emissions of NO_x, which is the pollutant of highest importance for these impact categories. Therefore, the upstream emissions cannot be neglected in general.

In this regard, the site-dependent impact assessment of the health effects of the upstream emissions, which are dominated by nitrate and sulfate aerosols, proved to be a worthwhile undertaking: The NO_x emissions for the supply of Diesel are about 1,6 times as high as those for the same amount of natural gas. Nevertheless, the associated health impacts are about equal for both fuels (tables 4.6 and 4.11). Likewise, the SO_2 emissions of the Diesel supply chain are about 6 times as high, but the associated health impacts only about 3 times as high as those for the supply of the same amount of natural gas (tables 4.7 and 4.12). This is because about 35 % of the NO_x and SO_2 emissions of the Diesel supply chain come from ships transporting oil from the OPEC countries to Germany. In the case of natural gas, on the other hand, a large contribution to the emissions (28 % for NO_x, 43 % for SO_2) comes from power plants in Germany delivering electricity for the compression of the gas.

The use of small natural gas fuel stations requiring a twofold compression energy compared to the large stations would increase the total impacts of the natural gas energy conversion chain by about 3 to 5 % in the case of buses and by between 5 and 20 % in the case of cars. The noticeable increase of 20 % thereby applies to the categories of acidification and human toxicity. This underscores the importance of making natural gas fuel stations efficient by connecting them to pipelines that provide high input pressures.

Possible future improvements of Diesel and petrol vehicles: The effect of the possible future development of the Diesel technology on the outcome of the comparison between Diesel and natural gas shall be discussed for the case of buses. According to table 4.18, the most significant future emission reductions for Diesel buses of a factor 3 for the EURO 4 compared to the EURO 2 emission control standard can be expected for NO_x and for particulate matter, which also represent the most important contributions to the overall impacts in the categories of acidification, nutrification, ozone formation and human toxicity. If no parallel reductions of the emissions of these pollutants from natural gas buses were achieved, the ratio of the impact scores of Diesel buses in these categories compared to natural gas buses could be reduced from about a factor of 5 to about a factor of 1,8 (table 5.7). However, if the lower NO_x emission factor of 1,0 g/km instead of 2,2 g/km from (Mangelsdorf et al. 1999, see table 4.18) was used to represent possible parallel

Table 5.7. Effect of possible future technological improvements of Diesel buses (EURO 4) and natural gas buses (lower NO_x emission factor) on the category indicator results

impact category	indicator	Diesel bus		natural gas bus	
		EURO 2	EURO 4	2,2 g NO_x / km	1,0 g NO_x/ km
resources	CED [MJ/km]	15,83	15,83	19,90	19,90
climate change	CO_2-eq. [g/km]	1194	1194	1210	1210
acidification	SO_2-eq. [g/km]	8,5	3,1	1,8	0,9
nutrification	N-eq. [g/km]	3,4	1,2	0,7	0,4
ozone formation	ethylene-eq. [g/km]	7,3	2,8	1,6	0,9
human toxicity	DALY [a/km]	1,2 E-6	3,8 E-7	1,9 E-7	1,0 E-7

CED Cumulative Energy Demand, *eq.* equivalents, *DALY* Disability Adjusted Life Years

improvements of the natural gas technology, relative disadvantages of Diesel (EURO 4) in the order of a factor 3 would remain with regards to these four categories. The impact scores with regards to resources and climate change would remain unchanged in all cases (table 5.7). Overall, it can be concluded that the relative disadvantages of the Diesel buses will become smaller with regards to several impact categories, but will not turn into advantages.

This also remains true if the use of particle filters for Diesel buses is considered, which could reduce their particle emissions to about 40 % of the EURO 4 value (Mangelsdorf et al. 1999:23). Due to the unchanged contribution of the nitrate aerosols, this would lead to a reduction of their impacts on human health to about 70 % of those for the EURO 4 buses without filters. Accordingly, the health impacts of Diesel buses with particle filters would still be higher than those of natural gas buses (with NO_x emissions of 2,2 g/km), but only by a factor 1,4, which may not be high enough to justify the introduction of a new fuel.

An analogous scenario analysis for the comparison of petrol and natural gas cars leads to similar results (table 5.8): the advantages of the natural gas vehicles with regards to acidification, nutrification, ozone formation and human toxicity would be reduced from a factor between 2,3 and 3,5 to a factor between 1,8 and 2,8 for petrol vehicles according to the EURO 4 emission control standard. The effect of the emission reductions of the petrol vehicles is smaller than in the case of the Diesel buses despite their similar relative size. This is due to the upstream emissions, which play a more important role for petrol than for Diesel and were assumed to remain unchanged. A parallel reduction of the NO_x emissions of the natural gas cars by 50 % could preserve the size of their relative advantage with regards to these four impact categories.

Emission factors: Due to the significant or even dominant contributions of the use phase to the overall impacts, the emission factors and fuel consumptions of the vehicles are important input parameters. Generally speaking, the fuel consumptions and the emission factors for regulated exhaust gas components (PM, NO_x, CO, VOC) are more reliable than the emission factors for the unregulated organic carcinogens, since a broader empirical basis is available for their determination.

Table 5.8. Effect of possible future technological improvements of petrol cars (EURO 4) and natural gas cars (lower NO_x emission factor) on the category indicator results

impact category	indicator	petrol car		natural gas car	
		EURO 2[a]	EURO 4[b]	0,06 g NO_x / km[c]	0,03 g NO_x/ km
resources	CED [MJ/km]	4,03	4,03	3,99	3,99
climate change	CO_2-eq. [g/km]	298	298	245	245
acidification	SO_2-eq. [g/km]	0,32	0,26	0,09	0,07
nutrification	N-eq. [g/km]	0,08	0,06	0,03	0,02
ozone formation	ethylene-eq. [g/km]	0,38	0,20	0,12	0,11
human toxicity	DALY [a/km]	2,6 E-8	1,9 E-8	8,5 E-9	6,2 E-9

CED Cumulative Energy Demand, *eq.* equivalents, *DALY* Disability Adjusted Life Years
[a] The emission factors from (Bach et al. 1998, see table 4.20) used here are inbetween the EURO 2 and EURO 3 emission factors from (Mangelsdorf et al. 1999:18).
[b] emission factors from (Mangelsdorf et al. 1999:18), in particular NO_x emissions reduced from 0,14 g/km for EURO 2 to 0,08 g/km; SO_2-emissions reduced to 25 % for petrol with 50 ppm sulfur content instead of 200 ppm for EURO 2; upstream emissions of 133 kg NMVOC/TJ petrol, which include evaporation from vehicles, reduced to 36 kg/TJ (see table 4.8, footnote b).
[c] value used here (Bach et al. 1998, see table 4.20).

Emission factors for the organic carcinogens, on the other hand, are either based on few measurements, sometimes close to the limits of detection, split factors that are applied to the total VOC emissions or analogies to other vehicle types. However, the regulated components PM and NO_x are those that contribute most to the human health effects. NO_x also generally dominates the impacts in the categories of acidification, nutrification and ozone formation, while the fuel consumptions of the vehicles by far dominate the impacts in the categories of resource extraction and climate change. The influence of the less reliable emission factors for the organic carcinogens on the overall results is therefore negligible.

For cars fueled by Diesel, petrol and natural gas, comparative test stand measurements for the *same* vehicle type referring to an idealized driving cycle were used. While these data may not exactly represent the absolute emissions under actual driving conditions, they nevertheless provide a good empirical basis for a *comparative* assessment. The emissions of the regulated components from Diesel buses were taken from a database that relies on extensive measurements (INFRAS 1999). The empirical basis for the emission factors of the regulated components for natural gas buses is weaker, such that the determination of more reliable emission factors is desirable. This applies in particular to the NO_x emissions, which dominate the human health effects. The full results of ongoing measurement projects, in particular the demonstration program of the German Federal Environmental Agency and the NGVeurope project were not available at the time of writing (September 1999). Therefore, a sensitivity analysis with regards to the emission factor for NO_x was carried out (see table 5.7).

Absolute size of the health impacts: The input parameters discussed above refer to properties of the energy conversion chains of natural gas, Diesel and petrol vehicles. They may affect the outcome of the comparison since they generally assume *different* values for the compared options. A further set of important input parameters and assumptions is related to the methodology used for the impact assessment in *all* cases. These parameters affect the *absolute size* of the impacts in the same way in each case.

Only the uncertainties associated with the new method for the assessment of the human health impacts proposed in chapter 3 will be considered, since the remaining impacts were assessed using standard methods from the literature. In this regard, the determination of the population exposures with pollutant dispersion models and the effect factors from the epidemiological literature need to be considered. Uncertainty estimates are provided in the following in terms of multiplicative confidence intervals, which are more suited to the typically lognormal distributions of the considered variables. The spread of a lognormal distribution with the geometric mean μ_g is characterized by the geometric standard deviation σ_g such that the 68 % confidence interval is $[\mu_g/\sigma_g, \mu_g \times \sigma_g]$ and the 95 % confidence interval is $[\mu_g/\sigma_g^2, \mu_g \times \sigma_g^2]$ (European Commission 1998:87-89; Hofstetter 1998: 424-426).

The uncertainty of the determination of the short-range contributions to the population exposures using a Gaussian dispersion model was estimated as $\sigma_g^2 = 2$ (section 2.5.2.5). A detailed estimate of the uncertainty of the Windrose Trajectory Model in terms of the determination of the long-range population exposures was not available, but due to the large integration area, the associated uncertainty is unlikely to be larger than in the case of the Gaussian dispersion model used for the short-range. Overall, the uncertainty in the determination of the population exposures is therefore estimated to be about a factor 2-3 for those pollutants dominating the human health effects, namely primary particles and secondary nitrate aerosols. This is consistent with an estimate of the uncertainty of the corresponding fate factors by Hofstetter (1998:311) as $\sigma_g^2 = 2$ for nitrates and $\sigma_g^2 = 3$ for particulate matter.

Much higher uncertainties are associated with the effect factors derived from epidemiological studies. The exposure response slopes for the dominant respiratory health effects of primary particulate matter (PM 10 and PM 2,5) are characterized by $\sigma_g^2 = 15$, and those for nitrate aerosols by $\sigma_g^2 = 34$, since they are set equal to those for the primary particles without direct epidemiological evidence (section 2.5.2.5). Together with an estimated uncertainty of $\sigma_g^2 = 3$ for the determination of the Years Lived Disabled (YLD) and the Years of Life Lost (YLL) per incidence of a respiratory disease and the uncertainty of the fate factors discussed above, Hofstetter (1998:338) attaches uncertainty estimates of $\sigma_g^2 = 19$ and $\sigma_g^2 = 36$ to the damage factors (YLL or YLD per mass of pollutant) of primary particles and nitrate aerosols, respectively. This means that the absolute health impacts of natural gas, Diesel and petrol vehicles, which are dominated by either primary particles or secondary nitrate aerosols, are uncertain by a factor 4 for Diesel cars or 6 otherwise in both directions with 68 % confidence, and by a factor 19 for Diesel cars or 36 otherwise with 95 % confidence.

Since the effect factors for nitrate aerosols are derived from those for primary particles, a change in the effect factors will affect all considered vehicles and fuels in the same way. The ratios between their impacts will therefore remain unchanged. This holds with the exception of the possibility that the transfer of the effect factors from primary particles to nitrates should turn out to be invalid, the effect of which on the comparison of natural gas and Diesel buses was discussed in section 5.1.1.3.

6 Summary and Outlook

A method was presented to calculate site-dependent impact indicators for human health effects of airborne pollutants suitable for application within Life Cycle Assessments. The method is applicable to primary air pollutants for which a linear exposure response function can be assumed at the population level and to secondary sulfate and nitrate aerosols. The linearity assumption represents a limitation, but nevertheless covers the most relevant primary air pollutants emitted from energy generation and surface transportation (particulate matter, organic carcinogens, nitrogen oxide, sulfur dioxide).

The method characterizes the population density around the emission source in terms of two discrete variables, namely the country and a settlement structure class within the country. The emission height is kept as a continuously varying parameter. The exposure of the population to a pollutant due to a given emission is calculated as an average for all sites within a settlement structure class. For this purpose, a verified statistical reasoning is used which is the main idea underlying the method. Due to the averaging of impacts within the settlement structure classes, the only meteorological input variable that needs to be considered is the annual mean wind speed.

The health impacts of traffic emissions of Diesel particles with an atmospheric residence time of 5 days were found to vary between large cities in agglomerations and rural districts by a factor 2,2 across Germany and by a factor 8 across Europe. For primary pollutants with short atmospheric residence times in the order of a few hours (such as formaldehyde and acetaldehyde), the variation between large cities and rural districts amounts to a factor 5 for Germany and a factor 50 for Europe. Most of the variation is due to the population density distribution around the source. The influence of the annual mean wind speed is smaller. The spatial differentiation decreases significantly with increasing emission height.

The impacts of secondary sulfate and nitrate aerosols, which are formed from emissions of sulfur dioxide and nitrogen oxide, respectively, are less sensitive to the emission height and the settlement structure class due to the time required for their formation. They are more sensitive to variations of the population density at a larger geographical scale. It is therefore sufficient to characterize the impacts of sulfate and nitrate aerosols by country average values. Across Europe, these vary by about a factor 30 for nitrates and about a factor 10 for sulfates.

Further research with regards to the method is required in order to extend the definition of the settlement structure classes and the calculation of the associated average impacts to other countries. In order to consider toxic air pollutants with nonlinear exposure-response functions, the settlement structures introduced here need to be combined with a classification of emission sites in terms of the surrounding background concentrations of the pollutants. The spatial differentiation of the impacts of ozone formed from nitrogen oxides and volatile organic compounds needs to be investigated.

The method presented here was applied within a Life Cycle Assessment of city buses and cars fueled by natural gas. Diesel buses and cars as well as petrol cars according to the EURO 2 emission control standard were considered as reference technologies. Both the vehicle emissions and the emissions associated with the extraction, processing and transport of the fuels were considered. The fuel consumption of the natural gas buses and cars were assumed to be higher than those of their Diesel counterparts by 25 % and 14 %, respectively.

Due to strong emission reductions, especially of nitrogen oxides and particulate matter, the impacts of the natural gas vehicles in the categories acidification, nutrification, ozone formation and human toxicity are typically about a factor 5 lower than those of the Diesel vehicles, while the impacts of the petrol cars are inbetween. In terms of climate change, the impacts of natural gas and Diesel vehicles are practically equal. The climate impacts of petrol cars are somewhat higher than those for Diesel and natural gas cars. This is mainly due to the high amount of energy required for the refining of petrol.

The only disadvantage of natural gas vehicles compared to Diesel vehicles occurs with regards to the extraction of abiotic resources due to the higher fuel consumption. However, in order for this disadvantage to shift the overall preference to the side of Diesel vehicles, this impact category would need to be attributed an unrealistically high weight. The overall comparison between natural gas cars and petrol cars is more straightforward: natural gas fares equal or better with regards to all impact categories. Therefore, it can be concluded for both buses and cars that natural gas is the most favorable fuel with regards to the considered impact categories.

In order to make the various types of considered carcinogenic and respiratory health effects comparable, they were expressed in terms of the Disability Adjusted Life Years, a human health indicator that has been developed under the auspices of the World Health Organization. Accordingly, the main contributions to the health impact of the vehicles come from secondary nitrate aerosols and, in the case of Diesel vehicles, also from primary fine particles. The health impacts of the upstream emissions associated with the extraction, processing and transport of the fuels are dominated by secondary nitrate and sulfate aerosols. Their share of the total health impacts is below or equal to 10 % in the case of buses and Diesel cars. In the case of petrol and natural gas cars, however, the upstream emissions are responsible for more than 40 % of the total health impacts. Therefore, the health effects of the upstream emissions cannot be neglected a priori. In particular, their relative importance may increase with future reductions of the vehicle emissions.

The total health impacts of the natural gas buses are lower by a factor 6,3 than those of the Diesel buses in large cities in agglomerated regions and by a factor 5,5 in small cities in Germany. Therefore, the substitution of Diesel by natural gas is almost as worthwhile in small cities as it is in large cities within agglomerations. This is due to the dominant contribution of the nitrate aerosols to the overall health effects in both cases. The total health impacts of natural gas cars are smaller by a factor of 10 in small cities and by a factor of 14 in large cities in Germany compared to their Diesel counterparts. Accordingly, the substitution of Diesel cars by natural gas is more worthwhile in large cities in agglomerations than in small cities. This is due to the predominant contribution of the primary particles to the overall health effects of the Diesel buses. Nevertheless, the health benefits of the substitution of Diesel by natural gas are in the same order of magnitude for all types of cities in Germany. The health benefits of substituting petrol cars by natural gas cars are essentially independent of the size of the city, since the contributions of secondary nitrate aerosols dominate in either case.

The method presented here allowed to consider the site-dependence of the human health impacts of airborne emissions from both vehicles and the large number of upstream processes in a consistent way, i.e. with the same level of detail, and with reasonable effort. In similar applications, it may also save the higher effort of going through site-specific impact assessments for a large number of processes.

In this regard, it is suggested to apply the method within a Life Cycle Assessment comparing electric vehicles and fuel cell vehicles to vehicles with combustion engines using Diesel, petrol and natural gas. The electric and fuel cell vehicles are characterized by a shift of emissions from the vehicles to upstream processes such as power plants or the production of hydrogen or methanol. In the case of primary pollutants, the human health impacts per emitted mass will generally be lower for these upstream processes than for the vehicle emissions, in particular if the latter occur in large cities. The method presented here would allow to determine whether this has an effect on the overall results of the comparison.

Besides studies on transportation systems, the use of site-dependent indicators for human health impacts appears to be particularly relevant in Life Cycle Assessments of products or services where transportation of goods with Diesel vehicles (trucks or vans) contributes significantly to the overall emissions. This may for example be the case for certain agricultural products or beverages sold in recyclable containers (Jørgensen et al. 1996). Furthermore, it is suggested to use these indicators in order to calculate average impacts for emissions from specific sectors of industry, as it was demonstrated here for the example of refineries in Germany. These could be useful in connection with the industry averaged emission data that are frequently used in Life Cycle Assessment.

References

Ahbe S et al. (1990) Methodik für Ökobilanzen auf der Basis ökologischer Optimierung. Schriftenreihe Umwelt, Nr. 133. Bundesamt für Umwelt, Wald und Landschaft, Bern

Alcamo J et al. (1991, eds) The RAINS model of acidification. Science and Strategies in Europe. Executive Report 18. International Institute for Applied Systems Analysis, Laxenburg

Bach C et al. (1998) Wirkungsorientierte Bewertung von Automobilabgasen. Bericht Nr. 160'928'2. Eidgenössische Materialprüfungs- und Forschungsanstalt (EMPA), Dübendorf

Bächlin W et al. (1995) Vergleich verschiedener einfacher Modelle - Screening Modelle - zur Berechnung der Immissionsbelastung im Straßenraum durch. Kfz-spezifische Schadstoffe. Schriftenreihe Umweltplanung, Arbeits- und Umweltschutz, Heft 191. Hessische Landesanstalt für Umweltschutz, Wiesbaden

Baldwin S et al. (1990, eds) Quality of Life. Perspectives and Policies. Routledge, London

Barrett K, Berge E (1996) Transboundary Air Pollution in Europe. EMEP/MSC-W Report 1/96. Norwegian Meteorological Institute, Oslo

Bartosch R (1997) Augsburg's way to become the first European model city for natural gas vehicles. In: ENGVA, European Natural Gas Vehicles Association et al. (eds) Dissemination Workshop on the Results of Natural Gas Vehicles for European Cities. ENGVA, Amsterdam, pp 61-69

BBR, Bundesamt für Bauwesen und Raumordnung (1998a) Aktuelle Daten zur Entwicklung der Städte, Kreise und Gemeinden. Berichte des BBR, Band 1. Bundesamt für Bauwesen und Raumordnung, Bonn

BBR, Bundesamt für Bauwesen und Raumordnung (1998b) personal communication

Bemtgen et al. (1997) The Natural Gas Bus Project Berlin. European Commission, Directorate-General XVII, Brussels

Bliefert C (1995) Umweltchemie. 1. Nachdruck. VCH, Weinheim

Boguski TK et al. (1996) LCA Methodology. In: Curran 1996, pp 2.1-2.37

Bonnefous S, Despres A (1990) The Eurogrid-Paradox386 Software User's Guide. Centre d'Etudes Nucléaires de Fontenay-Aux-Roses, Institut de Protection et de Surete Nucléaire, CEA, Fontenay-Aux-Roses

Bröchler S et al. (1999, eds) Handbuch Technikfolgenabschätzung. edition sigma, Berlin

Brunk C et al. (1991) Value Assumptions in Risk Assessment. Wilfried Laurier University Press, Waterloo

Bull KR (1995) Critical Loads - Possibilities and Constraints. Water, Air and Soil Pollution 85:201-212

Carter W (1994) Development of ozone reactivity scales for volatile organic compounds. Journal of the Air and Waste Management Association 44:881-889

Christoffer J, Ulbricht-Eissing M (1989) Die bodennahen Windverhältnisse in der Bundesrepublik Deutschland. 2., vollständig neu bearbeitete Auflage. Berichte des Deutschen Wetterdienstes Nr. 147. Deutscher Wetterdienst, Offenbach

Consoli F et al. (1993) Guidelines for Life-Cycle Assessment: A 'Code of Practice'. SETAC, Society of Environmental Toxicology and Chemistry, Brussels

Cowan CE et al. (1995) The Multi-Media Fate Model: A Vital Tool for Predicting the Fate of Chemicals. SETAC Press, Pensacola

Curran MA (1996, ed) Environmental Life-Cycle Assessment. McGraw-Hill, New York

Curran MA (1999) Editorial: The Status of LCA in the USA. International Journal of Life Cycle Assessment 4(3):123-124

Dabberdt WF et al. (1973) Validation and applications of an urban diffusion model for vehicular pollutants. Atmospheric Environment 7:603-618

Dedikov JV et al. (1999) Estimating methane releases from natural gas production and transmission in Russia. Atmospheric Environment 33(20):3291-3299

den Boeft J et al. (1996) CAR International: a simple model to determine city street air quality. The Science of the Total Environment 189/190:321-326

Derwent RG, Nodop K (1986) Long-range transport and deposition of acidic nitrogen species in north-west Europe. Nature 324:356-358

Derwent et al. (1996) Photochemical Ozone Creation Potentials for a Large Number of Reactive Hydrocarbons under European Conditions. Atmospheric Environment 30(2):181-199

Derwent RG et al. (1988) On the nitrogen budget for the United Kingdom and north-west Europe. Q.J.R. Meteorol. Soc. 114:1127-1152

Derwent RG et al. (1989) An Intercomparison of Long-Term Atmospheric Transport Models: The Budgets of Acidifying Species for the Netherlands. Atmospheric Environment 23(9):1893-1909

Derwent RG et al. (1998) Photochemical ozone creation potentials for organic compounds in Northwest Europe calculated with a master chemical mechanism. Atmospheric Environment 32:2429-2441

Deutsche Shell AG (1998) Fakten und Argumente. Aktuelle Themen aus der Minaralöl-wirtschaft. Ausgabe Oktober 1998. Deutsche Shell AG, Hamburg

DGMK, Deutsche Wissenschaftliche Gesellschaft für Erdöl, Erdgas und Kohle (1992) Ansatz-punkte und Potentiale zur Minderung des Treibhauseffektes aus Sicht der fossilen Energie-träger. Forschungsbericht 448-2. DGMK, Hamburg

Eerens HC et al. (1993) The CAR model: the Dutch method to determine city street air quality. Atmospheric Environment 27B(4):389-399

EMEP/MSC-W, Cooperative Programme for Monitoring and Evaluation of the Long Range Transmission of Air Pollutants in Europe, Meteorological Synthesizing Centre - West (1999) Transboundary Acidifying Air Pollution in Europe. MSC-W Status Report 1998. Norwegian Meteorological Institute, Oslo

EPA, Environmental Protection Agency (United States), Office of Health and Environmental Assessment (1996) Integrated Risk Information System (IRIS). On-line Database, http://www.epa.gov/ngispgm3/irisCincinatti

European Commission, Directorate-General XII (1995) ExternE, Externalities of Energy, Vols. 1-5. Prepared by ETSU, UK and IER, D. Office for Official Publications of the European Communities, Luxembourg

European Commission, Directorate-General XII (1998) ExternE, Externalities of Energy. Methodology Annexes. http://externe.jrc.es

European Oil & Gas Yearbook (1999) Urban-Verlag, Hamburg, Wien

EUROSTAT, Statistical Office of the European Communities (1995) Regionen, Systematik der Gebietseinheiten für die Statistik NUTS. Office for Official Publications of the European Communities, Luxembourg

Frank W et al. (1995) Das Screeningverfahren STREET zur Beurteilung der Immissionssituation an innerstädtischen Straßen und Kreuzungen. In: VDI, Verein Deutscher Ingenieure-Gesell-schaft Fahrzeug- und Verkehrstechnik (ed) Emissionen des Straßenverkehrs - Immissionen in Ballungsgebieten. VDI Verlag, Düsseldorf

Friedrich R et al. (1998) External Costs of Transport. Forschungsbericht Band 46, IER, Institut für Energiewirtschaft und Rationelle Energieanwendung, Universität Stuttgart. IER, Stuttgart

Friedrich U, Schierbaum I (1997) Vergleich Immissionsmessung - Immissionsberechnung für 14 verkehrsnahe Meßpunkte im Land Brandenburg. Gefahrstoffe - Reinhaltung der Luft 57:55-59

Frischknecht R et al. (1996) Ökoinventare von Energiesystemen. Grundlagen für den ökolo-gischen Vergleich von Energiesystemen und den Einbezug von Energiesystemen in Ökobi-lanzen für die Schweiz. 3. Auflage. Bundesamt für Energiewirtschaft, Bern

Frischknecht R et al. (1998) Einstein's Lessons for Energy Accounting in LCA. International Journal of Life Cycle Assessment 3(5):266-272

Fritsche UR et al. (1994) Gesamt-Emissions-Modell Integrierter Systeme (GEMIS) Version 2.1: Aktualisierter und erweiterter Endbericht. Öko-Insitut, Darmstadt, Freiburg, Berlin

Gethmann CF (1999) Rationale Technikfolgenbeurteilung. In: Grundwald A (ed) Rationale Technikfolgenbeurteilung. Konzepte und methodische Grundlagen. Springer-Verlag, Berlin, Heidelberg, New York, pp 1-10

Gerth WP, Christoffer J (1994) Windkarten von Deutschland. Meteorologische Zeitschrift, Neue Folge, 3:67-77

Gieryn TF (1995) Boundaries of Science. In: Jasanoff S et al. (eds) Handbook of Science and Technology Studies. Sage Publications, Thousand Oaks, London, New Delhi, pp 393-443

Goedkoop M (1995) The Eco-Indicator 95. Weighting method for environmental effects that damage ecosystems or human health on a European scale. Contains 100 indicators for important materials and processes. Final Report. Report number 9523. Netherlands agency for energy and the environment, Utrecht

Goedkoop M et al. (1998) The Eco-Indicator 98 Explained. International Journal of Life Cycle Assessment 3(6):352-360

Grunwald A (1999) Technikfolgenabschätzung. Konzeptionen und Kritik. In: Grundwald A (ed) Rationale Technikfolgenbeurteilung. Konzepte und methodische Grundlagen. Springer-Verlag, Berlin, Heidelberg, New York, pp 11-27

Guinee J, Heijungs R (1993) A Proposal for the Classification of Toxic Substances within the Framework of Life Cycle Assessment of Products. Chemosphere 26(10):1925-44

Hassel D et al. (1995) Abgas-Emissionsfaktoren von Nutzfahrzeugen in der Bundesrepublik Deutschland für das Bezugsjahr 1990. TÜV Rheinland Sicherheit und Umweltschutz GmbH, Köln

Hauschild M, Wenzel H (1998) Environmental Assessment of Products, Volume 2 - Scientific Background. Chapman & Hall, London

Hedden K et al. (1994) Erdölprodukte / Raffinerien. Projekt Klimaverträgliche Energieversorgung in Baden-Württemberg. Arbeitsbericht Nr. 14. Akademie für Technikfolgenabschätzung in Baden Württemberg, Stuttgart

Heijungs R et al. (1992) Environmental Life Cycle Assessment of Products. Guide - October 1992. CML, Centre of Environmental Science, Leiden

Heijungs R (1995) Harmonization of Methods for Impact Assessment. Environmental Science & Pollution Research 2(4):217-24

Hendrickson CT et al. (1998) Use of Economic Input-Output Models for Environmental Life Cycle Assessment. Environmental Science & Technology April 1998

Hertwich EG, Pease WS (1998) ISO 14042 Restricts Use and Development of Impact Assessment. International Journal of Life Cycle Assessment 3(4):180-181

Hettelingh JP et al. (1995) The use of Critical Loads in Emission Reduction Agreements in Europe. Water, Air and Soil Pollution 85:2381-2388

Hofstetter P (1996) Time in Life Cycle Assessment. In: Braunschweig A et al. (eds) Developments in LCA Valuation. IWÖ-Diskussionsbeitrag Nr. 32. Institut für Wirtschaft und Ökologie an der Hochschule St. Gallen (IWÖ-HSG), St. Gallen, pp 98-121

Hofstetter P (1998) Perspectives in Life Cycle Impact Assessment. A Structured Approach to Combine Models of the Technosphere, Ecosphere and Valuesphere. Kluwer Academic Publishers, Dordrecht

Houghton JT et al. (1996) Climate Change 1995 - The Science of Climate Change. Contribution of Working Group I to the Second Assessment Report of the Intergovernmental Panel on Climate Change. Cambridge University Press, Cambridge

Huisinga R (1985) Technikfolgenbewertung. Serapion, Frankfurt a.M.

IANGV, International Association for Natural Gas Vehicles (1999) IANGV Newsletter 52, June 1999

IER, Institut für Energiewirtschaft und Rationelle Energieanwendung, Universität Stuttgart (1998) EcoSense software, version 2.0. IER, Stuttgart

IFEU, Institut für Energie- und Umweltforschung, Heidelberg (1999) personal communication

INFRAS (1998) Ökoprofile von Treibstoffen. Umwelt-Materialien Nr. 104. Bundesamt für Umwelt, Wald und Landschaft, Bern

INFRAS (1999) Handbuch Emissionsfaktoren des Strassenverkehrs. Version 1.2. INFRAS, Bern

ISO, International Standards Organization (1997) ISO 14040: Environmental management - Life cycle assessment - Principles and framework. German version. Beuth Verlag, Berlin

ISO, International Standards Organization (1998) ISO 14041: Environmental management - Life cycle assessment - Goal and scope definition and life cycle inventory analysis. German version. Beuth Verlag, Berlin

ISO, International Standards Organization (1999a) ISO/DIS 14042: Environmental management - Life cycle assessment - Life cycle impact assessment. German version. Beuth Verlag, Berlin

ISO, International Standards Organization (1999b) ISO/DIS 14043: Environmental management - Life cycle assessment - Life cycle interpretation. German version. Beuth Verlag, Berlin

Janicke L (1998) personal communication

Jasanoff S (1987) Contested Boundaries in Policy-Relevant Science. Social Studies of Science 17:195-230

Jasanoff S (1990) The Fifth Branch: Science Advisers as Policymakers. Harvard University Press, Cambridge MA, London

Jischa MF (1997) Das Leitbild Nachhaltigkeit und das Konzept Technikbewertung. Chemie Ingenieur Technik 69:1695-1703

Jischa MF (1999a) TA in der Wissenschaft. In: Bröchler et al. 1999, pp 333-342

Jischa MF (1999b) Technikfolgenabschätzung in Lehre und Forschung. In: Petermann T, Coenen R (eds) Technikfolgen-Abschätzung in Deutschland. Campus Verlag, Frankfurt a.M., New York, pp 165-195

Johnson WB et al. (1973) An urban diffusion simulation model for carbon monoxide. Journal of the Air Pollution Control Association 23(6):490-498

Jolliet O (1994) Critical Surface-Time: an Evaluation Method for LCA. In: Udo de Haes HA et al. (eds) Integrating Impact Assessment into LCA. Proceedings of the LCA symposium held at the Fourth SETAC-Europe Congress, 11-14 April 1994, The Free University Brussels, Belgium. SETAC, Society of Environmental Toxicology and Chemistry - Europe, Brussels, pp 133-142

Jolliet O et al. (1996) Impact Assessment of Human and Eco-Toxicity in Life Cycle Assessment. In: Udo de Haes 1996a , pp 49-61

Jolliet O, Crettaz P (1997) Fate Coefficients for the Toxicity Assessment of Air Pollutants. International Journal of Life Cycle Assessment 2(2):104-110

Jørgensen A-M et al. (1996) Transportation in LCA. A Comparative Evaluation of the Importance of Transport in Four LCAs. International Journal of Life Cycle Assessment 1(4):218-220

Karman CC, Reerink H (1997) Dynamic Assessment of the Ecological Risk of the Discharge of Produced Water from Oil and Gas Producing Platforms. In: Ale BJM et al. (eds) Book of Abstracts, International Conference Mapping Environmental Risks and Risk Comparison RISK 97. Amsterdam, 21-24 October 1997. RIVM, Rijksinstituut vor volksgezondheit en milieu, Bilthoven, pp 41-47

Keller M, Kessler S (1998) Gasbusse in Basel. Erfahrungsbericht. Umwelt-Materialien Nr. 103. Bundesamt für Umwelt, Wald und Landschaft, Bern

Klöpffer W (1996) Verhalten und Abbau von Umwelt-Chemikalien. Physikalisch-chemische Grundlagen. ecomed verlagsgesellschaft, Landsberg

Koch R (1995) Umweltchemikalien. Physikalisch-chemische Daten, Toxizitäten, Grenz- und Richtwerte, Umweltverhalten. Dritte Auflage. VCH Verlagsgesellschaft, Weinheim

Koch W, Windt H (1988) Untersuchungen zur Ausbreitung von Automobilabgasen im Nahbereich einer Autobahn. Der Niedersächsische Minister für Umwelt, Hannover

Kolke R (1998) Technische Optionen zur Verminderung der Verkehrsbelastungen: Brennstoff-zellenfahrzeuge. Texte 33/99. Umweltbundesamt, Berlin

Kreibich R (1999) Technikbewertung, Ökobilanzierung und Technikgestaltung. Kernbestandteile einer innovationsorientierten Umweltpolitik und Nachhaltigen Entwicklung. In: Bröchler et al. 1999, pp 813-35

Krewitt W (1996) Quantifizierung und Vergleich der Gesundheitsrisiken verschiedener Stromerzeugungssysteme. Forschungsberichte des Instituts für Energiewirtschaft und Rationelle Energieanwendung (IER), Universität Stuttgart, Band 33. IER, Stuttgart

Krewitt W et al. (1998) Application of the Impact Pathway Analysis in the Context of LCA. International Journal of Life Cycle Assessment 3(2):86-94

Krüger R et al. (1997) Alternative Kraftstoffe. Möglichkeiten zur Minderung der VOC-Emissionen im Straßenpersonenverkehr von Baden-Württemberg. ecomed Verlagsgesellschaft, Landsberg

Kuhler M et al. (1994) Comparison between measured and calculated concentrations of nitrogen oxides and ozone in the vicinity of a motorway. The Science of the Total Environment 146/147:387-394

LAI, Länderausschuß für Immissionsschutz (1992) Krebsrisiko durch Luftverunreinigungen. Entwicklung von "Beurteilungsmaßstäben für kanzerogene Luftverunreinigungen" im Auftrage der Umweltministerkonferenz. Ministerium für Umwelt, Raumordnung und Landwirtschaft des Landes Nordrhein-Westfalen, Düsseldorf

Ludwig B (1995) Methoden zur Modellbildung in der Technikbewertung. CUTEC-Schriftenreihe Nr. 18. Papierflieger, Clausthal-Zellerfeld

Mackay D (1991) Multimedia Environmental Models. The Fugacity Approach. Lewis, Chelsea

Maibach M et al. (1995) Ökoinventar Transporte. Grundlagen für den ökologischen Vergleich von Transportsystemen und den Einbezug von Transportsystemen in Ökobilanzen. INFRAS, Zürich

Manier G (1971) Untersuchungen über meteorologische Einflüsse auf die Ausbreitung von Schadgasen. Berichte des Deutschen Wetterdienstes Nr. 124. Deutscher Wetterdienst, Offenbach

Mangelsdorf I et al. (1999) Durchführung eines Risikovergleiches zwischen Dieselmotoremissionen und Ottomotoremissionen hinsichtlich ihrer kanzerogenen und nicht-kanzerogenen Wirkungen. Berichte des Umweltbundesamtes 2/99. Erich Schmidt Verlag, Berlin

Marheineke T et al. (1998) Application of a Hybrid-Approach to the Life Cycle Inventory Analysis of a Freight Transport Task. SAE paper 982201. Proceedings of the 1998 Total Life Cycle Conference, Graz, Austria, December 1-3, 1998. SAE, Society of Automotive Engineers, Warrendale, pp 291-298

Marquenie JM, Verburgh JJ (1997) Impact Assessment on Exploration Drilling in the Wadden Sea, the Netherlands. In: Ale BJM et al. (eds) Book of Abstracts, International Conference Mapping Environmental Risks and Risk Comparison RISK 97. Amsterdam, 21-24 October 1997. RIVM, Rijksinstituut vor volksgezondheit en milieu, Bilthoven, pp 365-368

Marsmann M et al. (1999) Letter to the Editor, in Reply to Hertwich & Pease, Int. J. LCA 3(4) 180-181, "ISO 14042 Restricts Use and Development of Impact Assessment". International Journal of Life Cycle Assessment 4(2):65

Mittelstraß J (1995, ed) Enzyklopädie Philosophie und Wissenschaftstheorie, Band 2: H-O. Korrigierter Nachdruck. Verlag J.B. Metzler, Stuttgart, Weimar

Moussiopoulos N et al. (1996) Ambient Air Quality, Pollutant Dispersion and Transport Models. European Environment Agency, Copenhagen

Murray CJL (1996) Rethinking DALYs. In: Murray and Lopez 1996, chapter 1

Murray CJL, Lopez AD (1996) The Global Burden of Disease. Volume 1. Harvard University Press, Boston

MWV, Mineralölwirtschaftsverband (1998) Mineralölzahlen 1998. MWV, Hamburg (http//:www.mineraloelwirtschaft.de)

Nigge K-M (1995, unpublished) Conditionally and Inherently Normative Issues in Risk Assessment. Paper for the course ES6599 within the Masters program in Environmental Studies, York University, Toronto

Nigge K-M (1998) A method for the site-dependent Life Cycle Impact Assessment of toxic air pollutants from traffic emissions. SAE paper 982181. SAE Transactions - Journal of Passenger Cars 107:2008-18

Peebles MWH (1992) Natural Gas Fundamentals. Shell International Gas Limited, London

Pelli T (1985) Berechnungsformeln von CO-Immissionen in Straßenschluchten mit und ohne Lücken. Staub Reinhaltung der Luft 45(7/8):347-352

Pilkington A et al. (1997) Health Effects in ExternE Transport: Assessment and Exposure-Response Functions. Draft July 1997. Institute of Occupational Medicine, Edinburgh

Posch M et al. (1997) Calculation and mapping of critical thresholds in Europe. Status report 1997. Coordination Center for Effects, National Institute of Public Health and the Environment, Bilthoven

Potting J, Hauschild M (1997) Predicted Environmental Impact and Expected Occurrence of Actual Environmental Impact. Part II: Spatial Differentiation in Life-Cycle Assessment via the Site-Dependent Characterisation of Environmental Impact from Emissions. International Journal of Life Cycle Assessment 2(4):209-216

Rabl A, Spadaro JV (1998) Estimates of real damage from air pollution: site dependence and simple impact indices for LCA. Abstracts Book, 8th Annual Meeting of SETAC-Europe, 14-18 April 1998, Bordeaux. Presentation 7A4/035. SETAC, Society of Environmental Toxicology and Chemistry, - Europe, pp 125-6

Rausch L et al. (1998) Gesamt-Emissions-Modell Integrierter Systeme (GEMIS) v. 3.08 (software). Öko-Institut, Darmstadt

Rijkeboer RC, Hendriksen P (1993) Regulated and Unregulated Exhaust Gas Components from LD Vehicles on Petrol, Diesel, LPG and CNG. Report Nö. 93.OR.VM.029/1/PHE/RR. May 1993. TNO, Delft

Rodt S et al. (1998) NGV Experiences in Germany. Proceedings of the International Conference and Exhibition for Natural Gas Vehicles 'NGV '98', 26-28 May 1998, Cologne, Germany. BGW, Bundesverband der Deutschen Gas- und Wasserwirtschaft, Bonn, pp 208-217

Ropohl G (1997) Methoden der Technikbewertung. In: Westphalen R (ed) Technikfolgenabschätzung als politische Aufgabe. 3., gänzlich revidierte, neu bearbeitete Auflage. R. Oldenbourg, München, pp 177-202

Ropohl G (1999) Methoden in der Praxis. In: Rapp F (ed) Aktualität der Technikbewertung. Erträge und Perspektiven der Richtlinie VDI 3780. VDI Report 29. Verein Deutscher Ingenieure (VDI), Düsseldorf, pp 33-40

Rösgen H-J et al. (1997) Das Erdgasbusprojekt Berlin. Technischer Abschlußbericht. Senatsverwaltung für Bauen, Wohnen und Verkehr, Berlin

Ruhrgas AG (1999) Branchenreport: Erdgas im Energiemarkt 1998. Ruhrgas AG, Essen (http://www.ruhrgas.de/deutsch/branch/branch1.htm)

Saur K et al. (1996) Life Cycle Assessment as an Engineering Tool in the Automotive Industry. International Journal of Life Cycle Assessment 1(1):15-21

Schmidt-Seiwert V (1997) Landkarten zum Vergleich der Regionen Westeuropas. In: Hradil S, Immerfall S (eds) Die westeuropäischen Gesellschaften im Vergleich. Leske + Buderich, Opladen, pp 603-28

Schüle M (1997a) Erdgas-Großtankstelle Hannover. 2 Jahre Betriebserfahrung. gwf - Gas, Erdgas 138(5):267-72

Schüle M (1997b) Technisch-wirtschaftliche Auslegungskriterien für Erdgastankstellen. gwf - Gas, Erdgas 138(1):1-7

Schweimer GW (1998) Tree-Structures and Networks in LCI Mapping. Mathematical Solutions. SAE paper 982226. Proceedings of the 1998 Total Life Cycle Conference, Graz, Austria, December 1-3, 1998. SAE, Society of Automotive Engineers, Warrendale, pp 505-11

Seidler R (1994) Gesetzgeberische Anforderungen zur Umsetzung des §40 Abs. 2 BImschG. In: KRdL, Kommission Reinhaltung der Luft im VDI und DIN (ed) Ausbreitung von Kfz-Emissionen. Schriftenreihe Band 21. KRdL, Düsseldorf

Seigneur C (1993) Multimedia modeling. In: Zanetti P (ed) Environmental Modeling, Vol. I. Computer Methods and Software for Simulating Environmental Pollution and its Adverse Effects. Computational Mechanics Publications, Southhampton, pp 225-272

Seinfeld JH, Pandis SN (1998) Atmospheric Chemistry and Physics. From Air Pollution to Climate Change. John Wiley & Sons, New York

Seppälä J (1999) Decision Analysis as a Tool for Life Cycle Impact Assessment. LCA Documents, Vol. 4. ecomed Verlagsgesellschaft, Landsberg

Simpson D (1992) Long-Period Modelling of Photochemical Oxidants in Europe: Model Calculations for July 1985. Atmospheric Environment 26A:1609-1634

Simpson D (1993) Photochemical Model Calculations over Europe for Two Extended Summer Periods: 1985 and 1989. Model Results and Comparison with Observations. Atmospheric Environment 27A(6):921-943

Simpson D et al. (1997) Photochemical Oxidant Modelling in Europe: Multi annual Modelling and Source-receptor Relationships. EMEP/MSC-W Report 3/97. Norwegian Meteorological Institute, Oslo

Spadaro JV, Rabl A (1999) Estimates of Real Damage from Air Pollution: Site Dependence and Simple Impact Indices for LCA. International Journal of Life Cycle Assessment 4(4):229-243

Stanners D, Bourdeau P (1995) Europe's Environment. The Dobris Assessment. Prepared for the European Commission and the European Environment Agency, Copenhagen. Office for Official Publications of the European Communities, Luxembourg

Steinmüller K (1999) Methoden der TA - ein Überblick. In: Bröchler et al. 1999, pp 655-667

TA Luft (1986) Erste Allgemeine Verwaltungsvorschrift zum Bundes-Immissionsschutzgesetz (Technische Anleitung zur Reinhaltung der Luft - TA Luft) vom 27. Februar 1986. GMBl., S.95, ber. 202

Task Force on Mapping (1996) Manual on Methodologies and Criteria for Mapping Critical Levels / Loads and Geographical Areas where they are Exceeded. Texte 71/96. Umweltbundesamt, Berlin

Troen I, Petersen EL (1989) European Wind Atlas. Risø National Laboratory, Roskilde, Denmark

Udo de Haes HA (1996a, ed) Towards a Methodology for Life Cycle Impact Assessment. SETAC, Society of Environmental Toxicology and Chemistry - Europe, Brussels

Udo de Haes HA (1996b) Discussion of General Principles and Guidelines for Practical Use. In: Udo de Haes 1996a, pp 7-30

Udo de Haes HA et al. (1999) Best Available Practice Regarding Impact Categories and Category Indicators in Life Cycle Impact Assessment. Background Document for the Second Working Group on Life Cycle Impact Assessment of SETAC-Europe (WIA-2). International Journal of Life Cycle Assessment 4(2):66-73 and 4(3):168-174

Udo de Haes HA, Jolliet O (1999) How Does ISO/DIS 14042 on Life Cycle Impact Assessment Accomodate Current Best Available Practice. International Journal of Life Cycle Assessment 4(2):75-80

van der Wielen A (1996) Risk assessment of substances: an activity that needs intelligence. In: Vollmer G et al. (eds) Risk Assessment - Theory and Practice. Office for Official Publications of the European Communities, Luxembourg

van Dop H (1985) Atmospheric Dispersion of Pollutants and Modelling of Air Pollution Dispersion. In: Hutzinger O (ed) The Handbook of Environmental Chemistry. Volume 4: Air Pollution. Springer-Verlag, Berlin, pp 108-47

van Jaarsveld JA, de Leeuw FAAM (1993) OPS: an operational atmospheric transport model for priority substances. Environmental Software 8:91-100

van Pul WAJ et al. (1998) The potential for long-range transboundary atmospheric transport. Chemosphere 37(1):113-41

VDI, Verein Deutscher Ingenieure (1985) Guideline 3782, Part 3, Dispersion of Air Pollutants in the Atmosphere - Determination of Plume Rise. VDI-Verlag, Düsseldorf

VDI, Verein Deutscher Ingenieure (1991) Guideline 3780, Technology Assessment, Concepts and Foundations. Beuth Verlag, Berlin

VDI, Verein Deutscher Ingenieure (1992) Guideline 3782, Part 1, Dispersion of Pollutants in the Atmosphere - Gaussian Dispersion Model for Air Quality Management. Beuth Verlag, Berlin

VDI, Verein Deutscher Ingenieure-Gesellschaft Energietechnik (1997) Guideline 4600, Cumulative Energy Demand. Terms, Definitions, Methods of Calculation. Beuth Verlag, Berlin

VDI, Verein Deutscher Ingenieure-Gesellschaft Energietechnik (1998) Jahrbuch 98. VDI-Verlag, Düsseldorf

Volkmar H (1999) Quantifizierung nachhaltiger motorisierter Mobilität. CUTEC-Schriftenreihe Nr. 33. Papierflieger, Clausthal-Zellerfeld

WHO, World Health Organization (1987) Air Quality Guidelines for Europe. WHO, Geneva

WHO, World Health Organization (1996a) Revised WHO Air Quality Guidelines for Europe 1996. European Centre for Environment and Health, Bilthoven

WHO, World Health Organization (1996b) Diesel Fuel and Exhaust Emissions. Environmental Health Criteria 171. WHO, Geneva

Yetergil Kiefer ZD (1997) Externe Kosten von Krebserkrankungen durch kanzerogene Luftschadstoffe: Eine Abschätzung für die Schweiz mit besonderer Berücksichtigung von Benzol, polyzyklischen aromatischen Kohlenwasserstoffen und Dieselrusspartikeln. Dissertation Nr. 12308. ETH, Eidgenössische Technische Hochschule, Zürich

Zahn U et al. (1996) Diercke Weltatlas. Westermann Schulbuchverlag, Braunschweig

Zanetti P (1990) Air Pollution Modeling. Computational Mechanics Publications, Southhampton

Zenger A (1998) Atmosphärische Ausbreitungsmodellierung. Grundlagen und Praxis. Springer-Verlag, Berlin

Zimmermann V (1993) Methodenprobleme des Technology Assessment. Institut für Technikfolgenabschätzung und Systemanalyse, Forschungszentrum Karlsruhe, Karlsruhe

Zittel W (1997) Untersuchung zum Kenntnisstand über Methanemissionen beim Export von Erdgas aus Rußland nach Deutschland. Studie im Auftrag der Ruhrgas AG. Ludwig-Bölkow-Systemtechnik GmbH, Ottobrunn

Appendix

Substance Data and Population Exposures for Emissions in Germany

Table A.1. Substance parameters used for the calculation of population exposures

substance	v_d [cm/s]	v_w [a] [cm/s]	k [s^{-1}]	τ_a [b] [h]	reference
PM 2,5 (d < 0,95 μm[c])	0,065	0,10	0	135	IER 1998
PM 10 (d = 0,95-4 μm[c])	0,25	0,78	0	22	IER 1998
PM (d = 4-10 μm)	0,71	0,78	0	15	IER 1998
PM (d = 10-20 μm)	1,32	0,78	0	11	IER 1998
PM (d > 20 μm)	6,70	0,78	0	3	IER 1998
SO$_2$	0,50	0,10	2,8 E-6	27	IER 1998
NO$_x$	0,166	0,024	1,2 E-5[d]	19	van Jaarsveld and de Leeuw 1993:96
benzene	0	0	1,3 E-6	216	Hofstetter 1998:427, 243
formaldehyde	0	0	4,6 E-5	6	Seinfeld and Pandis 1998:110
acetaldehyde	0	0	3,1 E-5	9	Seinfeld and Pandis 1998:110
benzo[a]pyrene	0	0	5,9 E-6	47	Hofstetter 1998:427, 243
1,3-butadiene	0	0	2,5 E-5	11	Hofstetter 1998:427, 243

v_d dry deposition velocity, v_w wet deposition velocity, k decay rate due to chemical reactions, τ_a atmospheric residence time, d particle diameter

[a] $v_w = w_r \times p$, with the scavenging ratio w_r and a mean precipitation rate p=1,78 mm/d for the modeling area taken from (IER 1998).

[b] $\tau_a = [(v_d+v_w)/H + k]^{-1}$, where H = 800 m is the mixing layer height used in the Windrose Trajectory Model.

[c] The attribution of the diameter ranges d < 0,95 μm to PM 2,5 and of d = 0,95-4 μm to PM 10 is made for the purposes of dispersion modeling only, i.e. it only refers only to the respective substance data in columns 1 to 4, but not to the physical properties of relevance to their human health effects.

[d] for the reaction NO$_2$+OH $\rightarrow$ HNO$_3$ within the Harwell Trajectory Model, which is similar to the Windrose Trajectory Model used here (Derwent et al. 1989:1897).

Table A.2. (Part 1) Population exposures PE/M [persons ($\mu g/m^3$) a/kg] per mass of pollutant emitted in Germany for primary pollutants, differentiated according to the effective emission height h and the settlement structure class (notation as defined in table 3.3) and for a default annual mean wind speed (at 10 meters altitude) of 3,5 m/s. D $av.$ country-average value for Germany

h[m]	settlement structure class									
	I,1	I,2	II,1	I,3	II,3	I,4&5	II,4&5	III,4	III,5	D av.
PM 2,5										
5	2,64	1,88	1,74	1,48	1,45	1,32	1,29	1,28	1,18	1,41
10	2,49	1,83	1,68	1,46	1,43	1,31	1,28	1,27	1,18	1,40
20	2,29	1,76	1,59	1,43	1,40	1,30	1,27	1,25	1,17	1,37
35	2,12	1,70	1,51	1,41	1,38	1,29	1,26	1,24	1,17	1,35
50	2,01	1,66	1,46	1,39	1,37	1,28	1,25	1,23	1,17	1,34
75	1,92	1,63	1,42	1,39	1,36	1,28	1,25	1,23	1,16	1,33
100	1,80	1,58	1,37	1,36	1,34	1,26	1,24	1,22	1,16	1,30
125	1,69	1,52	1,32	1,34	1,32	1,25	1,23	1,20	1,15	1,29
150	1,57	1,45	1,29	1,30	1,30	1,23	1,22	1,19	1,15	1,26
175	1,51	1,40	1,26	1,27	1,27	1,21	1,20	1,18	1,14	1,24
200	1,47	1,37	1,25	1,25	1,26	1,19	1,19	1,17	1,13	1,22
PM 10										
5	1,72	1,04	0,94	0,70	0,66	0,56	0,54	0,53	0,45	0,64
10	1,61	1,00	0,89	0,68	0,65	0,56	0,53	0,52	0,45	0,63
20	1,44	0,95	0,82	0,66	0,63	0,55	0,52	0,51	0,45	0,61
35	1,28	0,90	0,75	0,64	0,62	0,54	0,52	0,50	0,44	0,59
50	1,19	0,86	0,70	0,63	0,61	0,54	0,51	0,50	0,44	0,58
75	1,10	0,84	0,66	0,62	0,60	0,53	0,51	0,49	0,44	0,57
100	0,99	0,79	0,62	0,60	0,58	0,52	0,50	0,48	0,43	0,55
125	0,89	0,74	0,58	0,58	0,56	0,51	0,49	0,47	0,43	0,54
150	0,79	0,68	0,55	0,55	0,54	0,49	0,48	0,46	0,43	0,52
175	0,74	0,64	0,53	0,53	0,53	0,48	0,47	0,45	0,42	0,50
200	0,71	0,62	0,52	0,52	0,52	0,47	0,46	0,45	0,42	0,49
SO_2										
5	1,62	0,99	0,90	0,67	0,64	0,55	0,53	0,52	0,44	0,62
10	1,54	0,97	0,86	0,67	0,64	0,55	0,53	0,52	0,45	0,62
20	1,40	0,93	0,80	0,66	0,63	0,55	0,52	0,51	0,45	0,61
35	1,27	0,89	0,74	0,65	0,62	0,55	0,52	0,51	0,45	0,60
50	1,19	0,87	0,71	0,64	0,61	0,55	0,52	0,51	0,45	0,59
75	1,10	0,84	0,67	0,63	0,61	0,54	0,52	0,50	0,45	0,58
100	1,00	0,79	0,63	0,61	0,59	0,53	0,51	0,49	0,44	0,57
125	0,91	0,76	0,60	0,60	0,58	0,53	0,51	0,49	0,45	0,56
150	0,81	0,70	0,57	0,57	0,57	0,51	0,50	0,48	0,45	0,54
175	0,77	0,67	0,56	0,56	0,55	0,50	0,49	0,48	0,45	0,53
200	0,74	0,65	0,54	0,55	0,55	0,50	0,49	0,47	0,44	0,52

Table A.2. (Part 2) Population exposures PE/M [persons ($\mu g/m^3$) a/kg] per mass of pollutant emitted in Germany for primary pollutants, differentiated according to the effective emission height h and the settlement structure class (notation as defined in table 3.3) and for a default annual mean wind speed (at 10 meters altitude) of 3,5 m/s. D $av.$ country-average value for Germany

h[m]	settlement structure class									
	I,1	I,2	II,1	I,3	II,3	I,4&5	II,4&5	III,4	III,5	D av.
NO_x										
5	1,89	1,17	1,04	0,79	0,75	0,64	0,61	0,59	0,51	0,72
10	1,76	1,13	0,98	0,77	0,74	0,63	0,60	0,59	0,50	0,71
20	1,58	1,07	0,90	0,75	0,72	0,62	0,59	0,58	0,50	0,69
35	1,41	1,01	0,82	0,73	0,70	0,61	0,58	0,56	0,49	0,67
50	1,31	0,97	0,78	0,71	0,69	0,60	0,58	0,56	0,49	0,66
75	1,22	0,94	0,74	0,70	0,68	0,60	0,57	0,55	0,49	0,65
100	1,10	0,89	0,69	0,68	0,65	0,58	0,56	0,54	0,48	0,62
125	1,00	0,84	0,64	0,66	0,64	0,57	0,55	0,53	0,48	0,61
150	0,88	0,76	0,61	0,62	0,62	0,55	0,54	0,52	0,48	0,58
175	0,83	0,72	0,59	0,59	0,59	0,53	0,52	0,50	0,47	0,56
200	0,79	0,69	0,57	0,57	0,58	0,52	0,51	0,49	0,46	0,55
benzene										
5	2,95	2,18	2,02	1,75	1,72	1,59	1,56	1,54	1,44	1,68
10	2,80	2,12	1,95	1,73	1,70	1,57	1,54	1,52	1,43	1,66
20	2,58	2,04	1,85	1,70	1,67	1,56	1,53	1,51	1,42	1,63
35	2,39	1,97	1,76	1,67	1,64	1,54	1,51	1,49	1,41	1,60
50	2,28	1,93	1,71	1,65	1,62	1,53	1,50	1,48	1,41	1,59
75	2,18	1,90	1,67	1,64	1,61	1,53	1,50	1,47	1,41	1,58
100	2,06	1,84	1,61	1,61	1,59	1,51	1,48	1,46	1,40	1,55
125	1,94	1,78	1,57	1,58	1,57	1,49	1,47	1,44	1,39	1,53
150	1,82	1,70	1,53	1,54	1,54	1,47	1,46	1,43	1,39	1,50
175	1,75	1,65	1,50	1,51	1,51	1,44	1,43	1,41	1,37	1,48
200	1,70	1,61	1,48	1,48	1,49	1,42	1,42	1,40	1,36	1,46
formaldehyde										
5	1,84	1,07	0,91	0,64	0,61	0,47	0,44	0,42	0,33	0,57
10	1,68	1,01	0,83	0,62	0,58	0,46	0,43	0,41	0,32	0,55
20	1,47	0,93	0,74	0,58	0,55	0,44	0,41	0,39	0,31	0,52
35	1,28	0,86	0,65	0,55	0,53	0,43	0,40	0,38	0,30	0,49
50	1,17	0,81	0,60	0,53	0,51	0,42	0,39	0,37	0,30	0,47
75	1,07	0,79	0,55	0,53	0,50	0,41	0,38	0,36	0,29	0,46
100	0,95	0,73	0,50	0,50	0,48	0,40	0,37	0,35	0,28	0,44
125	0,83	0,67	0,45	0,47	0,45	0,38	0,36	0,33	0,28	0,42
150	0,71	0,58	0,42	0,43	0,43	0,35	0,34	0,32	0,27	0,39
175	0,64	0,53	0,39	0,39	0,40	0,33	0,32	0,30	0,26	0,36
200	0,59	0,50	0,36	0,37	0,38	0,31	0,31	0,28	0,25	0,34

Table A.2. (Part 3) Population exposures PE/M [persons ($\mu g/m^3$) a/kg] per mass of pollutant emitted in Germany for primary pollutants, differentiated according to the effective emission height h and the settlement structure class (notation as defined in table 3.3) and for a default annual mean wind speed (at 10 meters altitude) of 3,5 m/s. D av. country-average value for Germany

h[m]	settlement structure class									
	I,1	I,2	II,1	I,3	II,3	I,4&5	II,4&5	III,4	III,5	D av.
acetaldehyde										
5	1,90	1,12	0,96	0,70	0,66	0,53	0,50	0,48	0,38	0,63
10	1,74	1,06	0,89	0,67	0,64	0,52	0,49	0,47	0,37	0,60
20	1,52	0,98	0,79	0,64	0,61	0,50	0,47	0,45	0,36	0,57
35	1,34	0,91	0,71	0,61	0,58	0,48	0,45	0,43	0,36	0,55
50	1,22	0,87	0,65	0,59	0,57	0,47	0,44	0,42	0,35	0,53
75	1,13	0,84	0,61	0,58	0,56	0,47	0,44	0,41	0,35	0,52
100	1,00	0,78	0,56	0,55	0,53	0,45	.0,42	0,40	0,34	0,50
125	0,89	0,72	0,51	0,52	0,51	0,43	0,41	0,39	0,33	0,47
150	0,76	0,64	0,47	0,48	0,48	0,41	0,40	0,37	0,33	0,45
175	0,69	0,59	0,44	0,45	0,45	0,38	0,38	0,35	0,31	0,42
200	0,65	0,55	0,42	0,42	0,43	0,37	0,36	0,34	0,31	0,40
B[a]P										
5	2,29	1,52	1,36	1,09	1,06	0,92	0,89	0,87	0,78	1,02
10	2,13	1,46	1,28	1,07	1,03	0,91	0,88	0,86	0,77	1,00
20	1,92	1,38	1,19	1,03	1,00	0,89	0,86	0,84	0,76	0,97
35	1,73	1,31	1,10	1,00	0,98	0,88	0,85	0,83	0,75	0,94
50	1,62	1,26	1,05	0,98	0,96	0,87	0,84	0,82	0,75	0,92
75	1,52	1,24	1,00	0,98	0,95	0,86	0,83	0,81	0,74	0,91
100	1,40	1,18	0,95	0,95	0,93	0,85	0,82	0,80	0,73	0,89
125	1,28	1,12	0,90	0,92	0,90	0,83	0,81	0,78	0,73	0,87
150	1,16	1,03	0,87	0,88	0,88	0,80	0,79	0,77	0,72	0,84
175	1,09	0,98	0,84	0,84	0,85	0,78	0,77	0,75	0,71	0,81
200	1,04	0,95	0,81	0,82	0,83	0,76	0,76	0,73	0,70	0,79
1,3-butadiene										
5	1,93	1,15	0,99	0,73	0,69	0,56	0,53	0,51	0,41	0,66
10	1,77	1,09	0,92	0,70	0,67	0,55	0,52	0,50	0,41	0,63
20	1,55	1,01	0,82	0,67	0,64	0,53	0,50	0,48	0,40	0,60
35	1,37	0,94	0,74	0,64	0,61	0,51	0,48	0,46	0,39	0,58
50	1,25	0,90	0,68	0,62	0,60	0,50	0,48	0,45	0,38	0,56
75	1,16	0,87	0,64	0,61	0,59	0,50	0,47	0,45	0,38	0,55
100	1,03	0,81	0,59	0,58	0,56	0,48	0,46	0,43	0,37	0,53
125	0,92	0,75	0,54	0,56	0,54	0,47	0,44	0,42	0,36	0,50
150	0,79	0,67	0,50	0,51	0,51	0,44	0,43	0,40	0,36	0,48
175	0,72	0,62	0,47	0,48	0,49	0,41	0,41	0,38	0,35	0,45
200	0,68	0,58	0,45	0,46	0,46	0,40	0,39	0,37	0,34	0,43

Table A.2. (Part 4) Population exposures PE/M [persons ($\mu g/m^3$) a/kg] to secondary pollutants per mass of primary pollutant emitted in Germany. *D av.* country-average value for Germany

	I,1	I,2	II,1	I,3	II,3	I,4&5	II,4&5	III,4	III,5	D av.
					settlement structure class					
sulfates from SO$_2$ all heights										0,145
nitrates from NO$_x$ all heights										0,240
ozone from NO$_x$ all heights										0,12
ozone from NMVOC all heights										0,12
ozone from CH$_4$ all heights										1,2 E-3

Acknowledgements

This dissertation was written while I was a research associate within the doctoral student program of the Europäische Akademie zur Erforschung von Folgen wissenschaftlich-technischer Entwicklungen GmbH. I wish to express my gratitude to the Europäische Akademie for its generous support of the dissertation.

I would like to thank Professor Dr.-Ing. Michael F. Jischa, Head of the Department of Fluid Dynamics and Systems Simulation at the Institute of Applied Mechanics of the Technical University of Clausthal for the supervision of the dissertation, and Professor Dr.-Ing. Hans-Peter Beck, Head of the Institute of Electrical Power Engineering of the Technical University of Clausthal for kindly accepting to be the second referee. I appreciate very much the critical and helpful discussions with both supervisors.

Furthermore, I gratefully acknowledge the constructive discussions in seminars with my colleagues at the Europäische Akademie and with my fellow doctoral students at the Institute for Technical Mechanics of the Technical University of Clausthal. In particular, I am grateful to Dipl.-Phys. Udo Lambrecht, Institute for Energy and Environmental Research (IFEU), Heidelberg, for valuable suggestions.

My special thanks go to Professor Dr.rer.nat. Franz Baumgartner, Interstate University of Applied Science, Buchs, Switzerland, Dipl.-Ing. Josef Decker, Clausthaler Umwelttechnik-Institut GmbH, Dr.rer.nat. Michael Decker, Europäische Akademie, Priv.-Doz. Dr.rer.nat. Armin Grunwald, Europäische Akademie (now Head of the Institute for Technology Assessment and Systems Analysis (ITAS) at the Karlsruhe Research Center), Dr.phil. Gerd Hanekamp, Europäische Akademie, and Dr.rer.nat. Stephan Saupe, German Aerospace Center (DLR), Cologne for critical reviews of parts of the manuscript and helpful suggestions for its improvement.

In der Reihe *Wissenschaftsethik und Technikfolgenbeurteilung*
sind bisher erschienen: